ALS

Advances in Life Sciences

Cytogenetics of Amphibians and Reptiles

Edited by
Ettore Olmo

1990

Birkhäuser Verlag
Basel · Boston · Berlin

Editor's address:

Prof. Ettore Olmo
Department of Comparative
and Evolutionary Biology
University of Naples
Via Mezzocannone, 8
80134 Naples
Italy

Cover illustration taken from A. Morescalchi's article (page 9, fig. 8).

Deutsche Bibliothek Cataloguing-in-Publication Data
Cytogenetics of amphibians and reptiles / ed. by Ettore Olmo. – Basel; Boston; Berlin: Birkhäuser, 1990
(Advances in life sciences)
ISBN 3-7643-2358-2 (Basel ...)
ISBN 0-8176-2358-2 (Boston)
NE: Olmo, Ettore [Hrsg.]

Printed in Germany on acid-free paper
ISBN 3-7643-2358-2
ISBN 0-8176-2358-2

TABLE OF CONTENTS

PREFACE

This book on the cytogenetics of amphibians and reptiles takes rise from a Symposium, having the same title, which was held in Canterbury (UK) on September 17th 1989, within the First World Congress of Herpetology. However, it does not only include the proceedings of the Symposium but also other interesting contributions.

Most of the articles included are review articles, and all the reports have been accepted only after a careful critical assessment.

The importance of a book on the cytogenetics of amphibians and reptiles derives from the increasing development that the research in this field has had in the last years. Moreover the herpetological cytogenetics has acquired a more and more relevant role both as a tool to study taxonomic and evolutionary problems and as a model to investigate more general problems, such as eukaryotic genome organization, genic regulation, the function and evolution of certain DNA fractions.

In my opinion, this multiplicity of aspects is well represented by the works included in this book, that provide a comprehensive overview of the present status of the research on cytogenetics of amphibians and reptiles.

Among taxonomic and evolutionary investigations, some papers are devoted specifically to certain taxa of amphibians and reptiles. Other are about more general problems, such as the role played by chromosome and DNA variation in the macroevolutionary and microevolutionary processes; the evolution of certain DNA fractions; the meaning of the so-called "C-value-paradox".

Several papers aim at improving the knowledge of chromosome composition, such as those using various banding techniques, those on the study of the origin, distribution and function of highly repetitive DNA fractions, those studying the localization of various types of proteins on lampbrush chromosomes by in situ hybridization with monoclonal antibodies, as well as the researches on the different mechanisms of sex chromosome differentiation.

Research concerning the relationships between chromosome and DNA variations and phenotype variations, mainly those affecting the developmental processes, has developed more recently, but is interesting all the same.

Mention should also be made of the increasing diffusion of research on the cytogenetics of amphibians and reptiles, which is witnessed by the different nationalities of the contributors and by the presence of scientists from countries like the People´s Republic of China, where so far the studies in this field had been very scarce.

This book is particularly addressed to all the students of cytogenetics and evolutionary biology of amphibians and reptiles, but I think it can be of great interest also for everyone involved in research on cytogenetics, genetics and zoology of vertebrates.

Ettore OLMO

LIST OF CONTRIBUTORS

B. ARANO. Museo Nacional de Ciencias Naturales de Madrid. C/ J.Gutierrez Abascal 2, 28006 Madrid, Spain.

R. BATISTIC. Servico de Genetica do Instituto Butantan, Av. Vital Brazil, 1500, Sao Paulo, SP, Brasil.

M.L. BECAK. Servico de Genetica do Instituto Butantan, Av. Vital Brazil, 1500, Sao Paulo, SP, Brasil.

W. BECAK. Servico de Genetica do Instituto Butantan, Av. Vital Brazil, 1500, Sao Paulo, SP, Brasil.

R.G. BELCHEVA. Department of Genetics, Faculty of Biology, University of Sofia, Bulgaria.

T. CAPRIGLIONE. Dipartimento di Biologia evolutiva e comparata. University of Naples. Via Mezzocannone 8, 80134, Naples, Italy.

A. CARDONE. Dipartimento di Biologia evolutiva e comparata. University of Naples. Via Mezzocannone 8, 80134, Naples, Italy.

M.L. ESTEBAN. Museo Nacional de Ciencias Naturales de Madrid. C/ J.Gutierrez Abascal 2, 28006 Madrid, Spain.

FEI LIANG. Chengdu Institute of Biology, Academia Sinica, Chengdu, Popular Republic of China.

B. FRITZ. Institut fur Genetik, Universitat Hohenheim, D-7000 Stuttgart, Federal republic of Germany.

J.P. FRITZ. Zoologisches Forschungsinstitut und Museum Alexander Koenig, Adenauerallee 150-164, D-5300 Bonn 1, Federal republic of Germany.

S. GARAGNA. Diparimento di Biologia animale. Piazza Botta 9. 27100 Pavia, Italy.

P. HERRERO. Museo Nacional de Ciencias Naturales de Madrid. C/ J.Gutierrez, Abascal 2, 28006 Madrid, Spain.

C. IKEBE. Department of Biology, Faculty of General Education, Toho University, Funabashi, Chiba 274, Japan.

M. KING. Museums of Arts and Sciences of the Northern Territory. G.P.O. Box 4646 Darwin, Australia 0801.

S. KOHNO. Department of Biology, Faculty of Sciences, Toho University; Funabashi, Chiba 274, Japan.

L. KUPRIYANOVA. Zoological Institute. The USSR Academy of Sciences 199034. Leningrad, U.R.S.S.

M. KURO-O. Department of Biology, Faculty of Sciences, Toho University; Funabashi, Chiba 274, Japan.

LI SHUSHEN. Department of Biology, Yunnan University, Kunming, Popular republic of China.

LI DEJUN. Department of Biology, Zunyi Medical College, Guizhou Province. Popular republic of China.

H.C. MACGREGOR. Department of Zoology, University of Leicester. Leicester LE1 7RH. England, Great Britain.

G.MANCINO. Dipartimento di Fisiologia e Biochimica. Istologia ed Embriologia, via Volta 4, 56100 Pisa, Italy.

D. MORAFKA. Department of Biology, California State University, Dominguez Hills, Carson, California 90747, U.S.A.

A. MORESCALCHI. Istituto di Anatomia comparata, the University, viale Benedetto XV 5, 16132 Genoa, Italy.

C. MORITZ. Department of Zoology, University of Queensland, St. Lucia QLD 4067, Australia.

G. ODIERNA. Dipartimento di Biologia evolutiva e comparata. University of Naples, via Mezzocannone 8, 80134 Naples, Italy.

E. OLMO. Dipartimento di Biologia evolutiva e comparata. University of Naples, via Mezzocannone 8, 80134 Naples, Italy.

M.N. RABELLO-GAY. Servico de Genetica do Instituto Butantan, Av. Vital Brazil 1500 Sao Paulo, SP, Brasil.

C.A. REDI. Dipartimento di Biologia Animale. Piazza Botta 9, 27100 Pavia, Italy.

M. SCHMID. Department of Human Genetics, University of Wurzburg, Koellikerstrasse 2, D-8700, Würzburg, Federal Republic of Germany.

Th.S. SOFIANIDOU. Department of Zoology, University of Thessaloniki, Greece

M. SOMA Servico de Genetica do Instituto Butantan, Av. Vital Brazil 1500, Sao Paulo, SP, Brasil.

S. STEWART. Department of Biology, California State University, Dominguez Hills, Carson, California 90747 U.S.A.

A. STOCK. Genetics Network, University of Nevada School of Medicine, 888 Willow St., Reno, Nevada 89502, U.S.A.

M. TORIBA. Japan Snake Institute, Yabuzuka-honmachi, Nitta-gun, Gunma, 379-23, Japan.

I. TRAJTENGERTZ. Servico de Genetica do Instituto Butantan, Av. Vital Brazil, 1500, Sao Paulo, SP, Brasil.

W. ULRICH. Institut fur Genetik, Universitat Hohenheim, D-7000 Stuttgart 70, Federal republic of Germany.

WEI GANG. Department of Biology, Zunyi Medical College, Guizhou Province, Popular republic of China.

XU NING. Department of Biology, Zunyi Medical College, Guizhou Province, Popular republic of China.

YE CHANGYUAN. Chengdu Institute of Biology, Academia Sinica, Chengdu, Popular republic of China.

PART 1. AMPHIBIA

Cytogenetics of Amphibians and Reptiles
Advances in Life Sciences

CYTOGENETICS AND THE PROBLEM OF LISSAMPHIBIAN RELATIONSHIPS

A. Morescalchi

Istituto di Anatomia comparata, the University, Viale Benedetto XV 5, 16132 Genoa, Italy.

SUMMARY: Genome evolution of the three orders of Lissamphibians seems to proceed from highly numbered karyotypes rich of microchromosomes to symmetrical numerically reduced karyotypes made up of only biarmed chromosomes: adaptive linkage-groups gained in evolution are thus tied together on few elements wich also contain multigene families which were localized before on microchromosomes. Some advanced groups of frogs exhibit acrocentric rich karyotypes probably deriving from symmetrical ones by means of centric fissions; their presence seems to parallel explosive speciation. Species showing similar reproductive or develepmental habits share comparable DNA amounts, notwithstanding their systematic position; thus changes in genome size are independent from phylogeny and related to metabolic or ontogenetic factors of adaptive value. Large genomes can be found in slowly developing species missing an aquatic larva or (in Urodeles) in groups heavily affected by paedomorphosis and paedogenesis. The different adaptive history of Urodeles on one hand, of Anurans and Apodans on the other, may give reason of the wide differences in genome size shown by members of the two groups.

INTRODUCTION.

Cytogenetics has largely improved our actual ideas on the natural history of living amphibians. Comparative karyology seems to support the notion of a common origin of the three orders (and the concept of Lissamphibia *sensu* Gadow) even if parallelisms in the quantitative differentiation of the nuclear DNA have been detected in many lineages of amphibians starting from paleozoic (Morescalchi, 1973, 1977; Thomson and Muraszko, 1978).

A first approach at the molecular level reveals that in its organization and composition the amphibian genome is more similar to that of fish than to that of reptiles (Bernardi *et al.*,

1985); moreover it is far more "plastic", quantitatively, than that of amniotes thus revealing peculiar properties easily interpreted as adaptive (Morescalchi and Olmo, 1982; Olmo et al,1989).

While the results coming from banding and other molecular techniques will be treated elsewhere in this Congress (on this point, see Schmid, 1980; Birnstein, 1982), I want to discuss the ossible phylogenetic role of gross chromosome and DNA changes detected in amphibian genomes. I will deal with the two subjects separately since in these animals the karyotype differentiation, in terms of chromosome numbers and shapes, seems to develop independently from that of the genome size (which influences only the absolute chromosome size).

THE EVOLUTION OF KARYOTYPE.

Most Lissamphibia show one of two karyotype models (Table I). The first, done by 42 or more elements (till 78 in diploid species), is called bimodal because it includes a good number of dot-like microchromosomes, more commonly found in primitive Osteoichtyes and in Sauropsids; this karyotype is also asymmetrical because formed by both biarmed (metacentric) and uniarmed (acrocentric) chromosomes. (Fig.1) The second model is unimodal and symmetrical because composed by only metacentric chromosomes; microchromosomes are missing and their peculiar DNA (rich of repetitive and multigene families) is now incorporated on biarmed chromosomes (see Vitelli et al., 1982). The diploid number is relatively low in symmetrical karyotypes: from 28 to 14 (Fig. 2); following Schmid (unpublished) a peculiar oriental ranid seems to show only 4 biarmed elements. Stebbins (1966) has deeply analised the genetic significance of symmetrical karyotypes, wich tie on few elements some groups of genes otherwise scattered in the genome; preservation of such linkage-groups con assure the maintenance of adaptive functions gained in previous evolutionary history.

If we look at the distribution of karyotype models in Lissamphibia it is easy to conclude that in the three orders the evolutionary advancement has been characterized by the gain of symmetrical formulae. Indeed, the most primitive families of Apodans (Ichthyophiidae) and of Urodeles (Hynobiidae and Cryptobranchidae) and the tailed-frog *Ascaphus* show bimodal karyotypes; genera of families considered "transitional" also from a phenotipic point of view (*Uraeotyphlus* and the old Caeciliid *Geotrypetes*; the Proteids and the Sirenids -though the last polyploid; some Discoglossids) show intermediate karyotypes; while the remaining families of Lissamphibia, generally considered "advanced" show (with some exceptions) symmetrical karyotypes.

Let us see some details of this process just to discuss the relationships of the various families.

Adopan karyology was reviewed recently by Nussbaum and Ducey (1988) who have described new seychellean material. Only 21 species belonging to 4 families are karyologically known: Ichthyophiidae show 42 chromosomes and a typical bimodal karyotypes and Seychellean and South American Caecilidae, with Typhlonectidae, show essentially symmetrical karyotypes. With minor variants, the trend in the karyotype evolution of this order is that previously suggested by Morescalchi (1973) and Nussbaum (1979) though data from the primitive Rhinatrematidae are necessary to fully confirm it.

Far more convincing is the picture on the order Urodeles: here nearly all living genera have been karyologically studied.

Following Schmalhausen (1968) *Ranodon* (family Hynobiidae)

TABLE I

CHROMOSOME NUMBERS (2n) IN AMPHIBIANS

APODANS:	Ichthyophiids: 42a
	Uraeotyphlids: 36a
	Caeciliids: 38a, 34a, 30a, 26t, 24
	Typhlonectids: 28, 24
URODELES:	Hynobiids: 78a, 66a, 62a, 58a, 56a, 40t
	Cryptobranchids: 60a
	Proteids: 38t
	Sirenids: 64P, 52P, 46P
	Amphiumids: 28
	Ambystomatids: 28, 26
	Plethodontids: 28, 26 (I)
	Salamandrids: 24, 22
ANURANS:	Discoglossoids: 46a, 38a, 28, 24, 18
	Pipoids: 24, 22, 20 (P,F)
	Pelobatoids: 26, 24
	Microhyloids: 28, 26, 24, 22
	Ranoids: 26, 24, 22, 20, 16, 14, 4 (F,P,I)
	Bufonoids: 26, 24, 22, 20, 19 (F,P,I)

a: asymmetrical bimodal karyotypes
t: transitional unimodal karyotypes
-: symmetrical unimodal karyotypes
F: increase of 2n by fissions
P: polyploidy
I: pericentric inversions

could be the most generalized living amphibian given by its structure, style of life and the crossopterygian-like larva: it has 66 chromosomes and a clearly bimodal karyotype (Fig. 3). All the remaining genera of this primitive family, with their paedogenetic derivatives Cryptobranchidae (Andrias and Cryptobranchus) (Fig. 4), show bimodal karyotypes and chromosome numbers between 78 (Onychodactylus) and 56 (Hynobius); only H. retardatus, a pond-living and relatively advanced species of the last genus, shows 40 chromosomes and a unimodal, transitional karyotype prefiguring that typical of Proteidae.

The perennibranchiate Proteids show 38 cromosomes, only one pair of which is dot-like; Sessions has discovered sex-chromosomes (male digamety) in some species of Necturus. Transitional karyotypes appear also in the Sirenidae (Fig. 5); the three living species seem to include populations with different degrees of ploidy, from tetraploidy on the way to diploidize (Siren intermedia, 46 chromosomes) to fully tetraploid and perhaps also exaploid animals (S.lacertina, 4n=52; Pseudobranchus striatus, 4n=64 or 48) (Morescalchi and Olmo, 1974; Morescalchi et al., 1986).

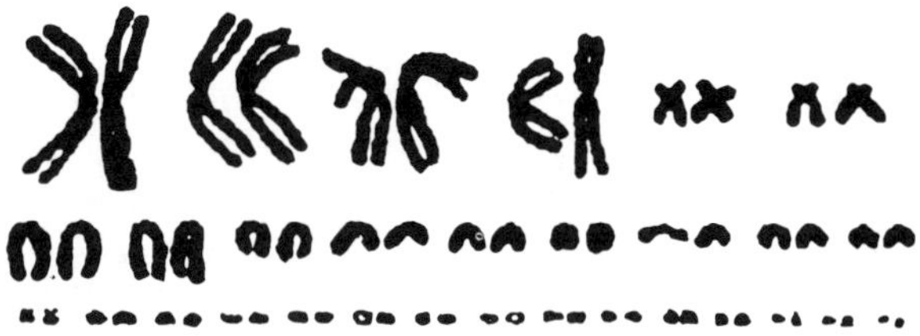

Fig. 1. The asymmetrical bimodal karyotype of a Cryptobranchid (2n=60).

Fig. 2. The symmetrical unimodal karyotype of a Myobatrachid (2n=24).

The only family of permanent neotenes showing an advanced karyotype is that of Amphiumidae, with 28 biarmed chromosomes which are similar, a part their higher absolute size due to the high DNA content of all neotenes, to those of most Ambystomatidae and the palearctic Plethodontidae (Fig. 6). In Rhyacotriton and in all Bolitoglossine plethodontids except Hydromantes (2n=28) this symmetrical karyotype is reduced to 26, usually biarmed,chromosomes (some pairs of the may be converted in acrocentric possibly through pericentric inversion, which happens in Aneides and in some Bolitoglossines; Kezer and Sessions, 1979). The Salamandridae show 24 biarmed chromosomes reduced to 22 in the more recent nearctic genera Notophtalmus and Taricha: also karyologically this family appears one of the advanced among Urodeles. Sex chromosomes have been described in some species of Salamandridae and they are particularly evident in some mesoamerican Bolitoglossine (see Morescalchi, 1975; Leon and Kezer, 1978; Schmid et al, 1979).

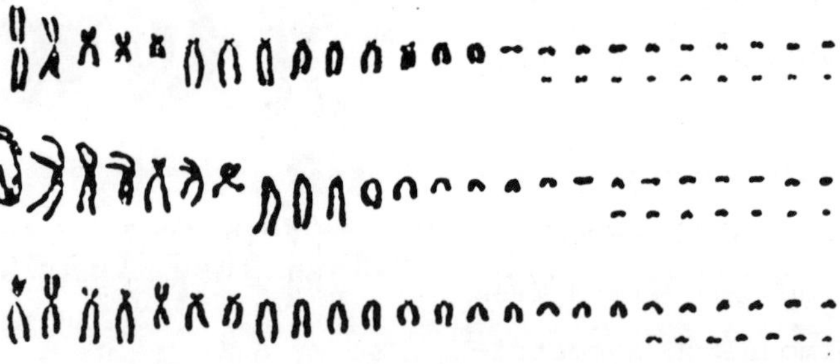

Fig.3. The haploid asymmetrical and bimodal karyotypes of the three Hynobiid genera, <u>Ranodon</u> (n=33), <u>Batrachuperus</u> (n=31) and <u>Salamandrella</u> (n=31)

The Urodeles offer a good example of "directionality" in chromosome evolution, with a progressive symmetrization along with phylogeny; events of speciation modify only the distribution of peculiar DNA families on chromosomes (i.e.,the nucleolar organizers, the 5S genes etc) but not the general karyotype morphology which in advanced groups appears "frozen" in standardized 28, 26, 24 and 22 chromosome-formulae.While the differentiation of sex-chromosomes or the rare cases of polyploidy allow specific departure from the standardized karyotypes. Also changes in the nuclear DNA amounts may parallel speciation.

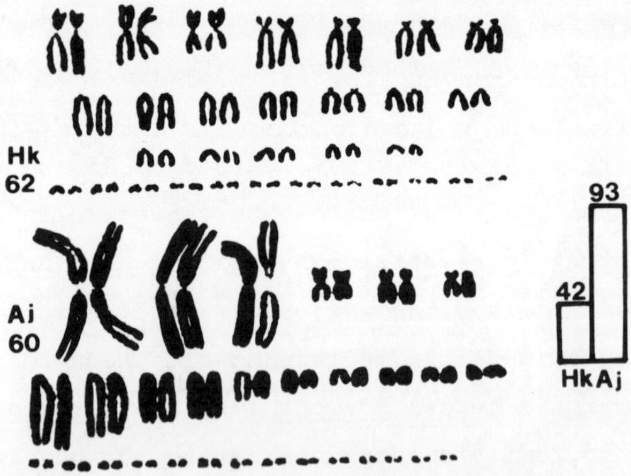

Fig. 4. The diploid sets of the Hynobiid <u>Salamandrella keyserlingii</u> (Hk) and of the Cryptobranchid <u>Adrias japonicus</u> (Aj); on the histogram the respective genome sizes (in picograms).

Far more complex is the picture in the Anurans, an order very rich of species most of which, especially in tropical habitats, are karyologically unknown (Fig. 7).

From a karyological viewpoint the erection of a superfamily Discoglossoids (Ascaphidae plus Discoglossidae) finds support through the fact that the karyotypes of Ascaphus (2n=46), Alytes (2n=38) and Discoglossus (2n=28) can be seriated along the trend from bimodal to symmetrical formulae; the last are fully reached by Bombina (2n=24) while Leiopelma seems to have exploited peculiar karyological routes having dot-like supernumerary chromosomes in one species and a reduced karyotype (2n=18) in two others (Stephenson et al., 1974). The Pipoids (Pipidae and Rhynophrynidae) show reduced chromosome numbers (from 24 to 20) in the diploid species, except Pipa parva which has 30 acrocentric chro-

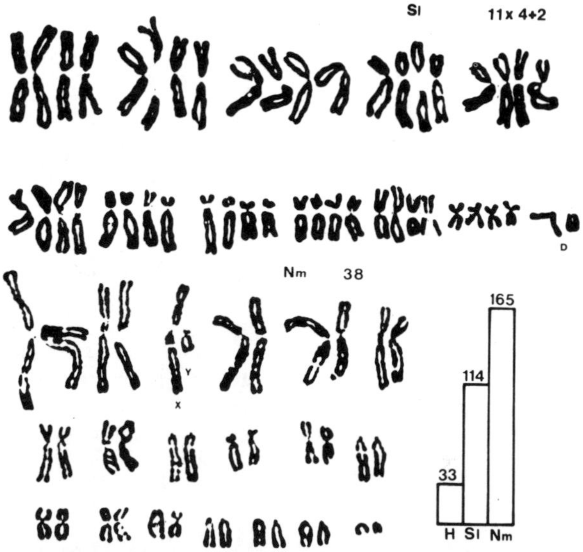

Fig. 5. The nearly tetraploid set of the Sirenid Siren intermedia (Si) and the diploid set of the Proteid Necturus maculosus (Nm, with sex chromosomes) on the histogram their genome size and that of a species of Hynobius (H).

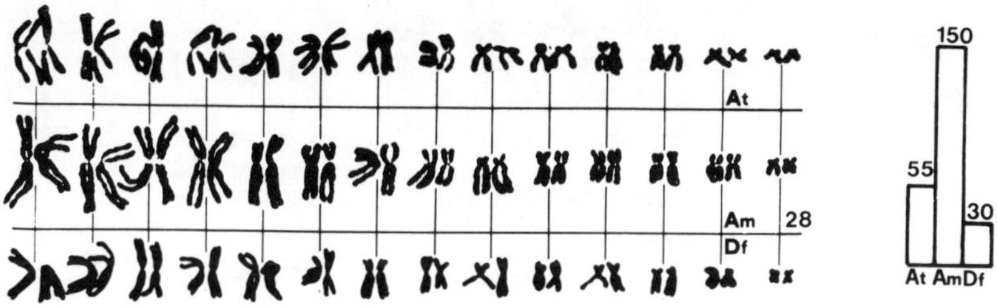

Fig. 6. The Ambystomatid Ambystoma tigrinum (At), the Amphiumid Amphiuma means (Am) and the Plethodontid Desmognathus fuscus (Df) have very similar symmetrical karyotypes (2n=28) but show largely different genome sizes.

mosomes possibly born by fission (Fig. 8); the genus *Xenopus* has widely exploited polyploidy with tetra-, exa- and octoploid species whose karyotypes are multiple of a *n*=10 or a *n*=9 basic set (Tymowska and Fischberg, 1982). Karyologically, Pipoids are nearer to the advanced than to the archaic frogs (Fig. 9).

Starting from Pelobatoids one can find symmetrical karyotypes formed by 26 chromosomes, reduced to 24 in some Megophryine and in *Pelodytes*. This 26-chromosome formula has been considered basal because, with small variants, it is present in many families of Microhyloids, Bufonoids and Ranoids. Inside each family the basal karyotype has apparently produced many times sets of 24, 22 or less chromosomes and in a parallel way in different subfamilies or genera; reduction has been consistent in African Petropedetine and Arthroleptine (2n=16, 14).

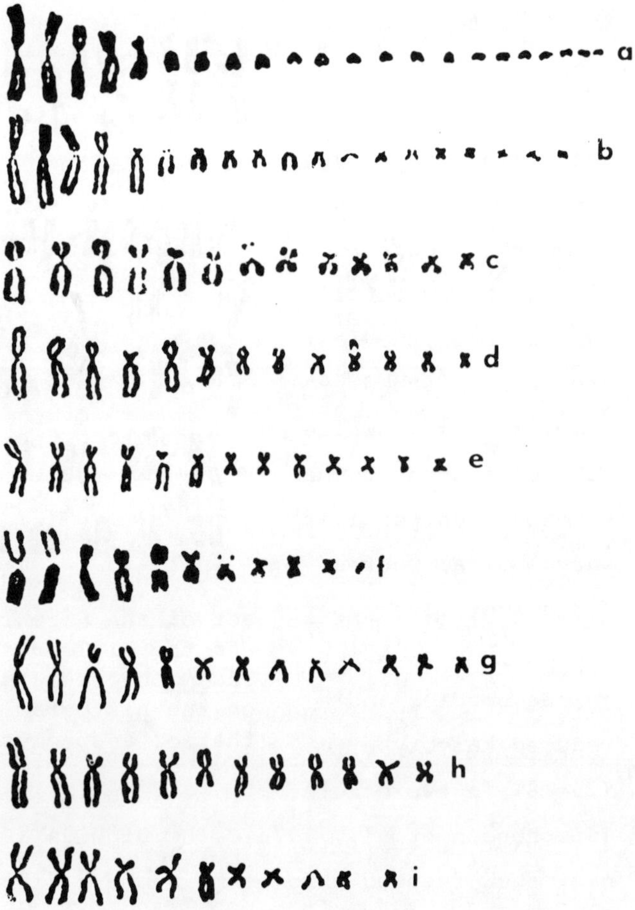

Fig. 7. Haploid karyotypes of representatives of the main anuran families (except Pipoids). a:Ascaphids; b:Discoglossids; c:Pelobatids; d:Leptodactylids; e:Hylids; f:Bufonids; g:Ranids; h:Hyperoliids; i:Microhylids.

Fig. 8. *Pipa parva* (Pipidae) has 30 acrocentric chromosomes possibly derived from centric fissions of metacentric elements.

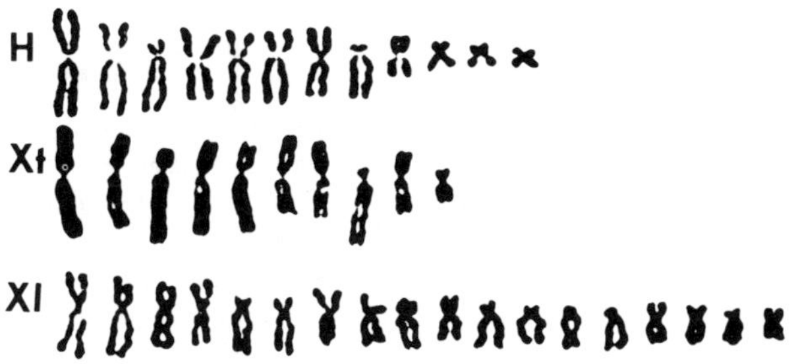

Fig. 9. Haploid sets of three Pipids: H: *Hymenochirus boettgeri* (2n=24); Xt: *Xenopus tropicalis* (2n=20); Xl:*Xenopus laevis* (2n=36; this karyotype is the product of polyplidization and shows "pairs" of homologous).

On the other hand,Jim Bogart (1973, 1981) has shown that some groups of tropical Ranoids and Bufonoids exhibit karyotypes rich in acrocentrics whose number increases in parallel to the diploid number (when 2*n*=36 usually all chromosomes are acrocentric). Sometimes these advanced groups preserve members with symmetrical karyotypes,which makes likely the hypothesis that acrocentrics have derived from metacentrics by means of centric fission.

Fissioning increases the chromosome number but doesn't radically change the composition of the main linkage-groups; it is thus suitable to create meiotic isolation as a premise for speciation. Indeed, in tropical habitats large groups of neobatrachians, and especially those showing acrocentrics, seem to be engaged in active (sometimes explosive) speciation.

Bogart argued on correlations between the acrocentric-rich karyotypes and the peculiar reproductive habits of tropical frogs, which are mostly terrestrial breeders and have direct development. Such correlations are not always valid (*Eupsophus* and *Pipa* have acrocentrics) but the absence of a free larval stage can offer the chance to evade selective constraints which in other frogs stabilize the karyotype morphology. In turn, this could favour speciation.

Other cytogenetic mechanisms having the same role are the differentiation of sex- chromosomes and polyploidy, which occur in some groups of Anurans without any apparent relations with their phylogenetic position (see Bogart, 1980; Mahony and Robinson, 1980; Schmid et al, 1988).

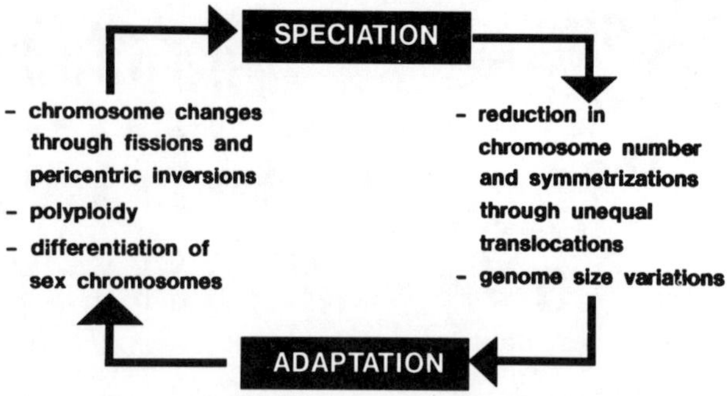

Fig.10. Cytogenetic mechanisms leading either to speciation or to adaptation seem to be rather distinct in Lissamphibians.

THE EVOLUTION OF GENOME SIZE.

The living amphibians exhibit a wide range of variability when nuclear DNA content is measured in the different species (Table II and Fig. 11; see Olmo, 1983); variability in genome size seems unrelated to karyotype morphology and doesn't depend on the systematic position (Fig. 4 - 6 ; Morescalchi, 1973, 1975). Some Urodeles (Amphiuma, Necturus) have more than 140 pg of DNA and show the largest genomes of all diploid vertebrates (only tetraploid lungfish exceed this amount). However, in Anurans genome size is lower reaching at best 40 pg, and comparable values are found in the few studied Apodans. Again in Anu

TABLE II

GENOME SIZE (2C: pg/n) IN AMPHIBIANS

APODANS:		from 7.4 to 26 (or more)
URODELES:	Hynobiids:	33-51
	*Cryptobranchids:	92-112
	*Proteids:	97-182
	*Sirenids :	90-114 (P)
	*Amphiumids:	130-150
	+Ambystomatids:	48-120
	Plethodontids:	30-86
	Bolitoglossines:	84-103
	Salamandrids:	39-86
ANURANS:		from 1.9 to 40

*: paedogenetic families
+: including Dicamptodontids

rans, species belonging to the same family (Myobatrachidae) show from 2 to 40 pg of DNA, with a difference of nearly 20 times in genome size which was detected in frogs, such as *Arenophryne* and *Limnodynastes*, living in similar, harsh environments but showing nearly opposite reproductive strategies (Mahony, Phil. D. Thesis, 1986).

Many hypotheses have been made on the meaning of the interspecific variability in genome size (see Cavalier-Smith,1985). Some Authors consider it nonsense because it is mainly due to fluctuations in junk or parasitic sequences deprived of any genetic role (John and Miklos, 1988); others consider it important in adaptive processes at cell or developmental level (Bennett, 1973).

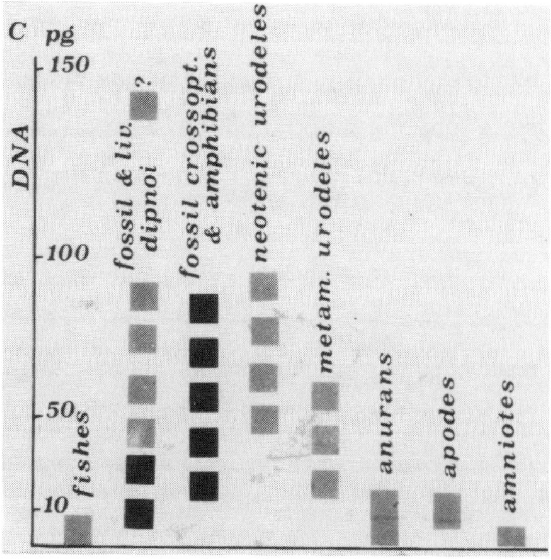

Fig. 11. The haploid (C) nuclear DNA amount in representatives of living (grey) and fossil (black) Vertebrates.

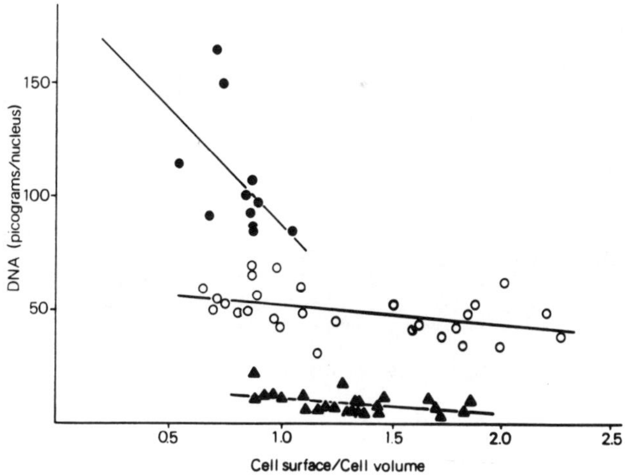

Fig.12. Relationships between the ratio of cell serface area to cell volume and the genome size (2C DNA) in Anurans (triangles),metamorphosing (open circles) and permanently neotenic Utodeles (solid circles)(Olmo and Morescalchi, 1978).

The adaptive hypothesis is based on the existence of correlations between genome size on one hand and cell size, nuclear size, cell metabolic rate, lenght of S-phase on the other.

Olmo and Morescalchi (1978) have found significant correlations between cellular parameters in a rather large sample of frogs and newts (Fig. 12); Olmo et al. (1978) and Macgregor (1982) have described correlations, though less precise, between genome size and the lenght of developmental processes in some species of amphibians. However, assumptions from the above reports have been at least partially cryticized by Larson (1984).

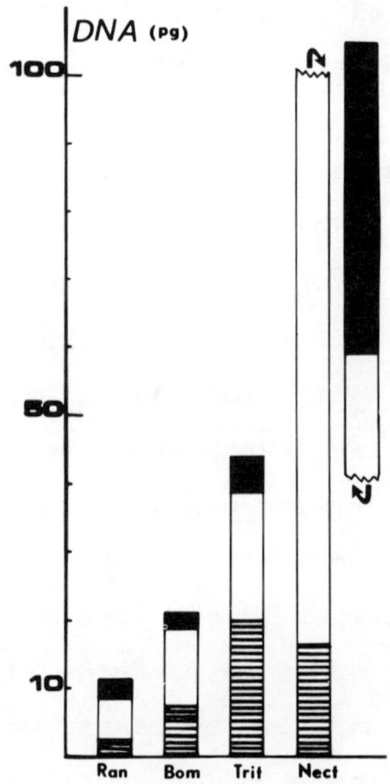

Fig.13. Amounts of single-copy (striped), middle repetitive (white) and highly repetitive DNA (black) in blood red cell nuclei of the Anurans Rana esculenta (Ran), Bombina variegata (Bom) and in the Urodeles Triturus cristatus (Tri), Necturus maculosus (Nect). (from data by Olmo, 1983).

Increase or decrease in the nuclear DNA amount involves all genomic fractions but especially the repetitive ones, perhaps as a result of the continuous turnover characterizing eukaryotic DNA; this process is potentially orthogenetic if selectively neutral (Dover, 1982).

In amphibians changes in genome size mostly involve satellite and middle repetitive DNA (Fig.13) (Macgregor et al., 1976; Olmo, 1983); however, far from being random, they are related to metabolic and ontogenetic factors which are of crucial importance in the adaptive strategies of these animals (Fig. 14). In Urodeles the highest DNA amounts are typical of species belonging to four unrelated families of permanently neotenic forms; also groups largely affected by paedomorphosis and low rates of metabolism (Rhyacotriton, neotropical Bolitoglossines) show very large genomes.

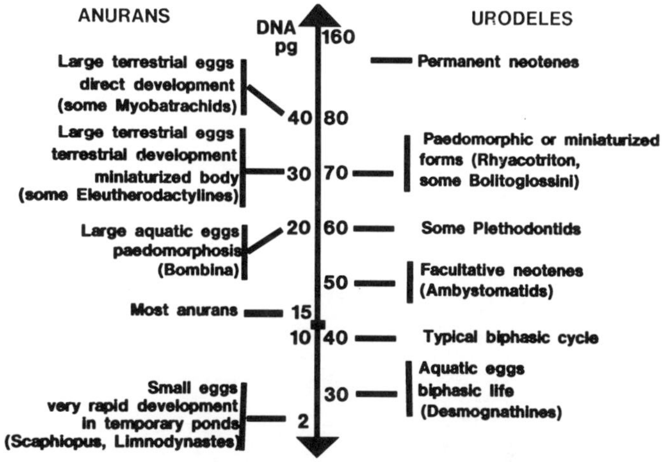

Fig. 14. Nuclear DNA contents and reproductive strategies in Anurans and Urodeles.

In Anurans the deserticoulous species which lay small eggs capable of rapid development in temporary waters (Scaphiopus, Limnodynastes) converge in showing very small genomes while the largest genomes in the order can be found in species showing a relented, often direct development from large eggs (Eleutherodactylus, Arenophryne, Myobatrachus). Thus it seems probable that the econiche peculiar of each amphibian species provides more or less severe metabolic and developmental constraints acting on genome size, which could explain why species sharing similar life history also show comparable DNA amounts (Morescalchi, 1975, 1980). Such conclusion allows a dichotomic interpretation of cytogenetic reports.

All lissamphibians show a very poor variation in their body structure; in each of the three orders, only small mioosteological differences do in fact permit the familial attribution of genera and species, often they are debated since parallelisms and convergences are very frequent in the differentiation of phenotypic attributes of these animals. (Inger, 1967; Edwards, 1976). At a karyological level, the limited anagenetic progress inside each order finds some correspondence in the scarce variation in

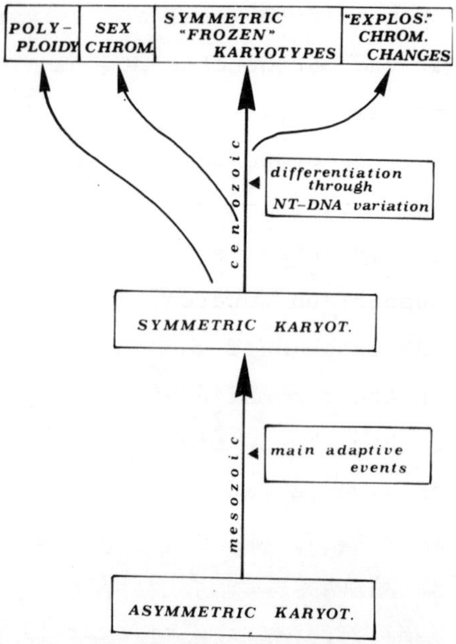

Fig.15. Probably the Lissamphibians gained symmetrical karyotypes at the end of mesozoic after which their karyological differentiation was achieved mainly through variations in the repetitive fractions of nuclear DNA.

chromosomal linkage-groups shown by the higher species, mostly showing "frozen" symmetrical karyotypes (Fig.15); (see also Wilson et al., 1974).On the other hand,lissamphibians are extremely plastic in their physiology, which allows them to occupy a number of divesified econiches also in side the same habitat (Savage, 1973, Duellman and Trueb, 1986). In late paleozoic, the amphibians were displaced by early reptiles and they survived only in cryptic or marginal microabitats in which their competitors could not enter (Fig.16) (Schmalhausen,1968).

Here the Apodans and the Urodeles -except the neotropical Plethodontids- were subject to limited radiation while the Anurans, starting from the cretaceous, gave rise to many advanced groups (neobatrachians) which still today are radiating in the tropics. The divergence of extant species seems to be largely due to metabolic adaptations rea-

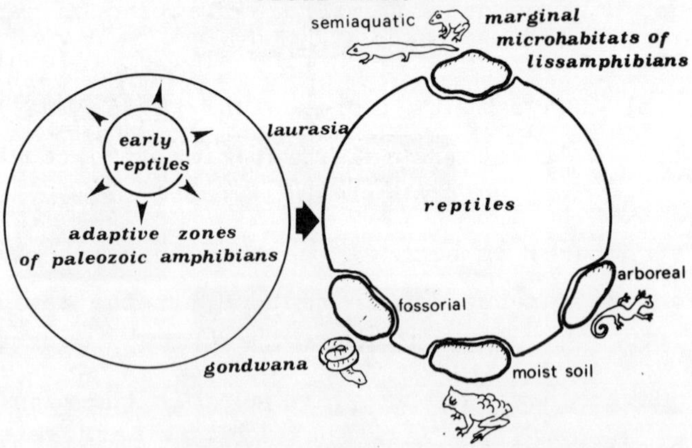

Fig.16. Radiation of Reptiles in late paleozoic has confined lissamphibians in cryptic or marginal microhabitats.

ched at cell level and to perturbations in time (heterochrony) of the ontogenetic process leading to phenotypes capable of exploiting new environmental situations. Heterochronic phenomena, such as progenesis and neoteny, lead to paedomorphic phenotypes and paedomorphosis is likely to have characterized the evolution of amphibians since their first appearance as fossil records (Hecht and Edwards, 1977; Bolt, 1977). While heterochrony seems responsable of the survival and radiation of most lissamphibians in their peculiar microhabitats (Wake, 1966; Wake et al., 1983).

Both cell metabolism and heterochrony seem to be heavily influenced by genome size (Morescalchi, 1975;Olmo and Morescalchi, 1975, 1978; Macgregor and Sessions, 1986). Indeed, the large cells of Urodeles require energetic budgets significantly lower (25%) than those of Anurans (Feder, 1977) and the lowest metabolic levels are typical of permanent neotenes and of lungless salamanders, which are endowed with hypertrophic genomes. Sessions and Larson (1987) have discovered inverse relationships between genome size and the differentiation rate of Plethodontids, arguing that changes in genome size may be important correlates of

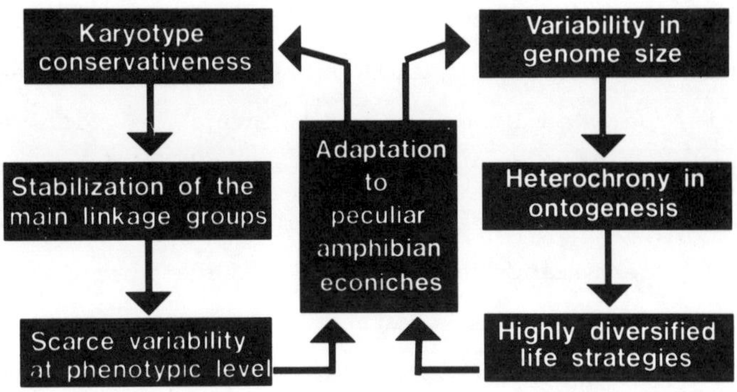

Fig.17. Adaptation of advanced lissamphibians is paralleled by genome size modulations while their karyotype is relatively conservative in the structure of linkage-groups.

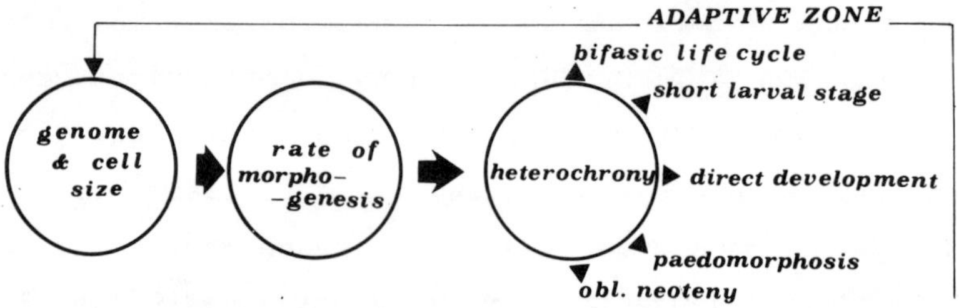

Fig. 18. In each adaptive zone the available energy budget selectively influences the cell (and genome) size and the last, in turn, the morphogenetic rate which can produce heterochrony and thus novel phenotypes or new reproductive strategies.

heterochronic mechanisms which repattern urodelan morphology in evolution (Fig.17). Research in this field is not so advances in Anurans, but here again the evolutionary role of heterochronic processes seems to be decisive (Trueb and Alberch, 1985) while the genera most involved in them do show high DNA amounts (Bombina, the miniaturized Eleutherodactyline, the paedomorphic Myobatrachids).

In conclusion the interspecific variability in genome size appears as the cytological mirror of changes in physiology and ontogeny which are at the basis of amphibian adaptation and differentiation (Fig.18). In this sense the large differences of genome size between Urodeles and Anurans may find a historical interpretation.

During the permotriassic, a midlatitude arid belt may have isolated prosalamanders in high latitude humid zones of future laurasia and proanurans (with proapodans) in the equatorial humid zones (Fig. 19) (see Duellman and Trueb, 1986).

Urodele evolution in mountain stream waters was accomplished under selective pressures different from those acting on the remaining amphibians and favouring the rise of large cells with low metabolic rates (the "frugal strategy" by Szarski, 1983).

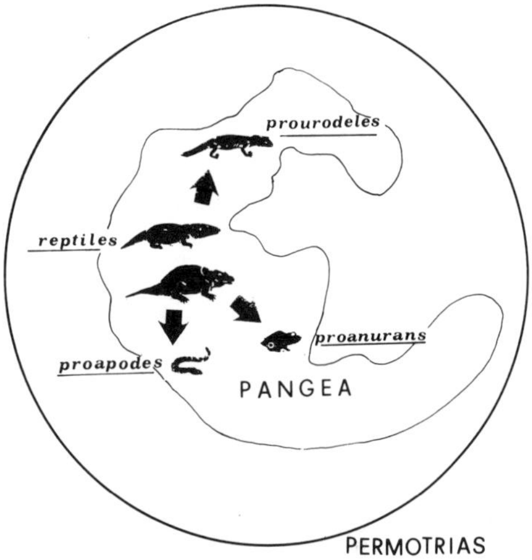

Fig.19. Reptiles or permotriassic climatic events have displaced prourodeles in northern territories of pangea and the precursors of the other lissamphibians in southern (gondwanan) territories.

In turn, large cells promote long division periods and can influence ontogeny producing paedomorphosis and sometimes paedogenesis; paedogenetic phenotypes adapted to harsh conditions have been positively selected at least four times in history of the order giving rise to the Cryptobranchids, Proteids, Sirendis and Amphiumids, all provided with enormous DNA quantitaties. Working on fossil material, Thomson and Muraszko (1978) have shown that large cells have arisen many times and in many lineages in the course of amphibian and lungfish phylogeny (Fig. 11); probably their history is that of the periodical environmental crises which occured from the paleozoic to recent times, driving in a parallel manner the differentiation of genome size both of the paedogenetic Urodeles and of the living Dipnoi (Morescalchi, 1975, 1977; Cavalier-Smith,1985).

REFERENCES.

Bennett, M.D. (1973). Brookhaven Symp.Biol. 25, 344-366.
Bernardi, G., Olofsson, B., Filipski, J., Zerial, M., Salinas, J., Cuny, G., Meunier-Rotival, M. and Rodier, F. (1985). Science 228,953-958.
Birnstein, V.J. (1982). J.Mol.Evol. 18. 73-91.
Bogart, J.P. (1973). In: Evolutionary biology of the Anurans. (J.L. Vial Ed.), Univ.Missouri Press, Columbia, Missouri, pp.337-349.
Bogart, J.P. (1980). In: Polyploidy, biological relevance. (W.H.Lewis Ed.), Plenum Press, N.Y., pp. 341-378.
Bogart, J.P. (1981). Monit.Zool.Ital.(N.S.), Suppl. 15:29-40.
Bolt, J.R. (1977). J.Paleontol. 51: 235-249.
Cavalier-Smith, T. (1985). In: The evolution genome size. (T.Cavalier-Smith Ed.), J.Wiley & Sons, Chichester-N.Y.-Brisbane-Toronto-Singapore, pp. 105-187.
Dover, G.A. (1982). Nature 229: 111-117.
Duellman, W.E. and Trueb, L. (1986). Biology of Amphibians. Mc Graw-Hill Co., N.Y.-St Louis-San Francisco.
Edwards, J.L. (1976). J. Morph. 148: 305-328.
Feder, M.E. (1977). J. Exp.Biol. 202: 403-414.
Hecht, M.K. and Edwards, J.L. (1977). In: Major patterns in vertebrate evolution. (M.K. Hecht, P., Goody and B.M. Hecht Eds.), Plenum Press, N.Y., pp. 3-51.
Inger, R.F. (1967). Evolution 21: 369-384.
John, B. and Miklos, G. (1988). The eukaryote genome in development and evolution. Allen & Unwin, London.
Kezer J. and Sessions, S.K. (1979). Chromosoma 71: 65-80.
Larson, A. (1984). Evol. Biol. 17: 119-217.
Leon, P.E. and Kezer, J. (1978). Chromosoma 65: 213-230.
Macgregor, H.C. (1982). In: Genome evolution (G.A. Dover and R.A. Flavell Eds.), Acad. Press, London, PP. 325-341.
Macgregor, H.C. and Sessions, S.K. (1986). Phil.Trans.R.Soc. Lond. B 312: 243-259.
Macgregor, H.C., Mizuno, S. and Vlad, M. (1976). In: Chromosomes Today (P.L. Pearson and K.R. Lewis Eds.), Vol.5, Wiley, N.Y., PP. 331-339.
Mahony, M.J. and Robinson, E.S. (1980). Chromosoma 81: 199-220.
Morescalchi, A. (1973). In: Cytotaxonomy and vertebrate evolution. (A.B. Chiarelli and E. Capanna Eds.), Acad.Press, London, pp. 233-348.
Morescalchi, A. (1977). In: Major patterns in vertebrate evolution. (M.K. Hecht, P. Goody and B.M. Hecht Eds.), Plenum Press, N.Y., pp. 149-167.

Morescalchi, A. (1980). Boll.Zool. (Suppl.) 47: 113-126.
Morescalchi, A. and Olmo, E. (1982). Cytogenet.Cell Genet. 34: 93-101.
Nussbaum, R.A. (1979). Occ.Pap.Mus.Zool.Univ.Michigan 687:1-20.
Nussbaum, R.A. and Ducey P.K. (1988). Herpetologica 44: 290-296.
Oeldorf, E., Nishioka, M. and Bachmann, K. (1978). Z.zool.Syst. Evolution.forsch. 16: 216-224.
Olmo, E. (1983). Bas.appl.Histochem. 27: 227-256.
Olmo, E. and Morescalchi, A. (1975). Experientis 31: 804-806.
Olmo, E. and Morescalchi, A. (1978). Experientis 34: 44-46.
Olmo, E. Capriglione, T. and Odierna, G. (1989). Comp.Biochem. Physiol. 92B: 447-453.
Savage, J.M. (1973). In: Evolutionary biology of the Anurans. (J.L.Vial Ed.)Univ.Missouri Press, Columbia,Missouri pp. 351-445.
Schmalhausen, I.I. (1968). The origin of terrestrial vertebrates. Acad. Press, N.Y.
Schmid, M. (1980). In: Wierbeltiere Zytogenetik, S.N.G.Jahrbuch (H.Muller Ed.), Birkhauser Verlag pp. 4-27.
Schmid, M., Olert, J. and Klett, C. (1979). Chromosoma 71:29-55.
Schmid, M. Steinlein, C., Feichtinger, W., de Almeida, C.G. and Duellman, W.E. (1988). Chromosoma 97: 33-42.
Sessions, S.K. and Larson, A. (1987). Evolution 41: 1239-1251.
Stebbins, G.L. (1966). Science 152: 1463-1469.
Stephenson, E.M., Robinson, E.S. and Stephenson, N.G. (1974). Experientia 30: 1248-1250.
Szarski, H. (1983). J.Theor.Biol. 105: 201-209.
Thomson, K.S. and Muraszko, K. (1978). J.Exp.Zool. 205: 315-320.
Trueb, L. and Alberch, P. (1985). Fortschr. Zool. 30: 113-121.
Tymowska, J. and Fischberg, M. (1982). Cytogenet.Cell. Genet. 34: 149-157.
Vitelli, L., Batistoni, R., Andronico, F., Nardi, I. and Barsacchi-Pilone, G. (1982). Chromosoma 84: 475-491.
Wake, D.B., (1966). Mem.S.Calif.Acad.Sci. 4: 1-111.
Wake, T.A., Wake, D.B. and Wake, M.H. (1983). J.Herpetol. 17: 10-22.
Wilson, A.C., Sarich, V.M. and Maxson, L.R. (1974). Proc.Nat. Acad.Sci.USA 71: 3028-3030.

Cytogenetics of Amphibians and Reptiles
Advances in Life Sciences

CHROMOSOME BANDING IN AMPHIBIA

M. Schmid[1], C. Steinlein[1], I. Nanda[1], and J.T. Epplen[2]

[1]Department of Human Genetics, University of Würzburg, Koellikerstr. 2, D-8700 Würzburg, Federal Republic of Germany

[2]Max-Planck-Institute for Psychiatry, Am Klopferspitz 18A, D-8033 Martinsried, Federal Republic of Germany

SUMMARY: The chromosomes of 400 species representing the main genera of Amphibia were analyzed with banding techniques. The purpose of this study was to recognize differences and common characteristics in the banding patterns found in amphibian chromosomes at various stages of evolution. In contrast to mammals and birds, but similar to many fish and reptiles, the euchromatic segments of amphibian chromosomes contain no multiple G-, Q- and R-bands. In most instances, amphibian chromosomes are distinguished by high amounts of constitutive heterochromatin. As revealed by DNA base pair specific fluorochromes, most of these heterochromatic regions are composed of GC-rich DNA. The nucleolus organizers in amphibian karyotypes are distinguished by a high degree of interindividual variability. Many of the nucleolus organizer polymorphisms are due to tandem duplications or triplications. The banding techniques made it possible to demonstrate the existence of primitive and highly heteromorphic XY- and ZW sex chromosomes in Amphibia.

The methods of chromosome analyses developed during the past two decades make it to date possible to perform very precise studies on the fine structure of vertebrate chromosomes. Chromosomal homologies across species and chromosomal anomalies associated with congenital malformations can be recognized using

G-, Q- and R-banding techniques. The constitutive heterochromatin can be differentially labeled by C-banding or staining with AT- or GC base pair specific fluorochromes. Furthermore, protocols have been developed which make it possible to visualize the time sequence of DNA replication in the chromosomes (BrdU-banding), or to specifically stain active nucleolus organizers or centromeres. Finally, the so-called "era of molecular cytogenetics" was spawned by the *in situ* hybridization of nucleic acids to fixed metaphase chromosomes. Complete manuals compiling all these methods have been published by Rooney & Czepulkowski (1986), Macgregor & Varley (1988), and Verma & Babu (1989).

The modern cytogenetic techniques were developed primarily on human chromosomes and later applied with great success to a large number of other mammals. Compared to the numerous reports dealing with banded mammalian chromosomes, only a few are completed on cold-blooded vertebrates (fish, amphibians and reptiles). However, it is to be expected that more valuable banding data on the chromosomes of these organisms will emerge in the near future.

GENERAL PROPERTIES OF CHROMOSOME BANDING IN AMPHIBIA

The analysis of chromosomal banding patterns with fluorochromes is performed by staining fixed and air-dried metaphases without any pretreatment, followed by the observation of these preparations under a fluorescence microscope equipped with appropriate optics and filter combinations. These DNA-specific fluorochromes induce characteristic transversal bands along the metaphase chromosomes of higher vertebrates (mammals and birds). The fluorochromes can be classified into two groups with complementary base pair specificity. The AT-specific fluorochromes such as quinacrine, DAPI, daunomycin, and Hoechst 33258 all produce Q-banding patterns that correspond to those revealed by the protein-denaturing G-banding procedures. The GC-specific fluorochromes like chromomycin, mithramycin and olivomycin, on the other hand, induce the complementary reverse (R-) banding patterns (for reviews see Latt, 1977; Comings, 1978). The fact that fluorochromes with complementary base pair specificity also produce complementary chromosome bands, constitutes strong indirect proof of the reciprocal AT- and GC base pair composition of Q- (or G) and R-bands (Schweizer, 1976; Schweizer et al., 1987; Comings & Drets, 1976; Latt et al., 1977).

In contrast to the warm-blooded mammals and birds, the chromosomes of amphibians, fishes and most reptiles contain no multiple fluorescent Q- and R-bands (Schmid 1980; Schmid & Guttenbach, 1988). Their euchromatic segments fluoresce with a uniform and moderate intensity (Fig. 1). All of those chromosome regions fluorescing either distinctly brighter or weaker than the euchromatin can be identified as constitutive heterochromatin by C-banding or as nucleolus organizers by silver staining (Fig. 2). It is also not posible to induce multiple G-band patterns by the enzymatic treatment of amphibian chromosomes with proteinases (Schmid, 1978).

The absence of multiple banding patterns in the euchromatin of amphibians was originally attributed to the strong spiralization of their chromosomes at metaphase (Schmid, 1978, 1980). Thus, comparative measurements between human chromosomes and those of ten toad species of the genus _Bufo_ showed that the

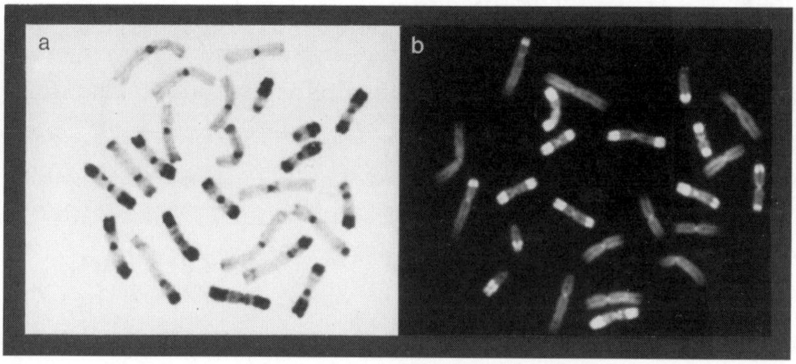

Fig.1. Metaphases of _Mixophyes fasciolatus_ after (a) C-banding and (b) staining with Hoechst 33258. Note in (b) the brightly fluorescing heterochromatin and the absence of banding patterns in the euchromatic chromosome segments.

average amount of DNA per µm of chromosome length was higher by a factor of 1.5 to 3 in the amphibian metaphases. It was concluded from this that the distances between individual bands in amphibian chromosomes are probably too small to be resolved (Schmid, 1978). However, later studies on other amphibian species showed that the DNA density in their metaphase chromosomes are as low as in mammals (i.e., 0.04 pg DNA/µm chromosome), or even 1.3 to 2 times less (Schmid et al., 1987). It was nevertheless not possible to induce multiple banding patterns in any of these species. Furthermore, reasonably distinct BrdU-replication patterns were obtained in the chromosomes of several

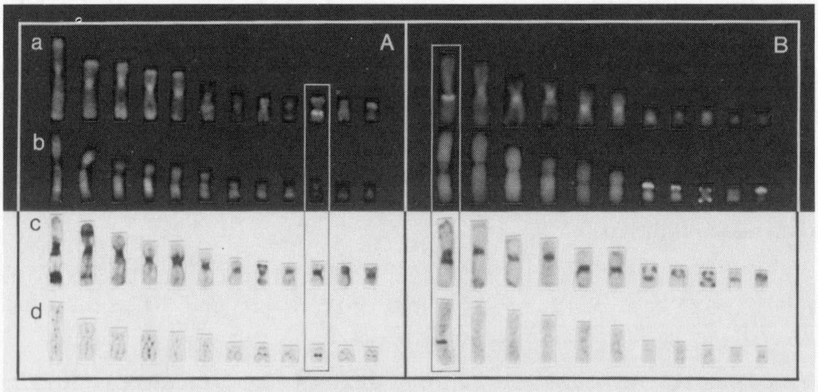

Fig.2. Karyotypes of (A) Rana macrodon and (B) Bufo americanus showing (a) mithramycin staining, (b) quinacrine fluorescence, (c) C-banding, and (d) silver staining. The NOR-carrying chromosomes are framed.

Anura and Urodela, regardless of their considerably different degree of spiralization (Schempp & Schmid, 1981; Kuro-o et al., 1986). All these data show that there must be other factors than the degree of chromosome spiralization responsible for the failure of attempts at euchromatin banding in Amphibia.

A generally accepted and attractive explanation for the presence of high quality G-, Q- and R-band patterns in metaphase chromosomes of warm-blooded vertebrates and their almost complete absence in the chromosomes of cold-blooded vertebrates is discussed in the following. Equilibrium density gradient centrifugation in the presence of certain sequence-specific ligands permitted the fractionation of the main-band DNA from warm-blooded vertebrates into light (AT) and heavy (GC) major components (Filipski et al., 1973; Thiery et al., 1976; Macaya et al., 1976). The DNA fragments from these major components originate during their separation by breakage from much longer DNA segments (≫300 kb) called "isochores" (Cuny et al., 1981), which have a rather homogeneous base pair composition and are distinguished by different GC levels. Furthermore, the evolutionary conservation of chromosome banding patterns in mammals is paralleled by conservation of the base pair composition and the relative amounts of the major components in genomes of the same class (Thiery et al., 1976; Cuny et al., 1981; Olofsson & Bernardi, 1983). The average amount of DNA per band in mammalian chromosomes at the limit of microscopic resolution is 1,250 kb (Holmquist, 1988), and is thus in the range of the isochores sizes. Finally, there are strong indications that G-bands correspond to AT-rich DNA and R-bands to GC-rich DNA (Schweizer, 1976; Latt et al., 1977; Comings, 1978). These results demonstrate that the genomes of the homoiothermic mammals and birds are strongly compartmentalized by base composition and clearly suggest that the isochores could correspond to the DNA sequences in G-, Q- and R-bands (Cuny et al., 1981). In contrast, isochores are either absent or rare in the genomes of the poikilothermic fishes, amphibians and -with reservations - of reptiles (Thiery et al., 1976; Hudson et al., 1980; Pizon et al., 1984; Medrano et al., 1988). This is paralleled by the difficulties in obtaining multiple euchromatin banding in their chromosomes. A full work-up of this correlation between DNA base composition heterogeneity and chromosomal banding would require supplementary density gradient centrifugation with the DNA of those reptiles whose chromosomes show different degrees of banding intensity.

CONSTITUTIVE HETEROCHROMATIN

Most amphibian karyotypes are characterized by many constitutively heterochromatic regions. Generally, the number of heterochromatic (C-) bands is largely independent of the length or the DNA content of the chromosomes (Schmid, 1978). Small chromosomes can contain just as much or even more heterochromatin as the largest chromosomes in a karyotype (Fig. 3). This holds true for highly evolved as well as for the more primitive amphibian families. In the chromosomes of the Anura, the centromeric/pericentromeric regions and the telomeres are the preferential locations of heterochromatin. Additionally, interstitial C-bands can be found. The nucleolus organizer regions are

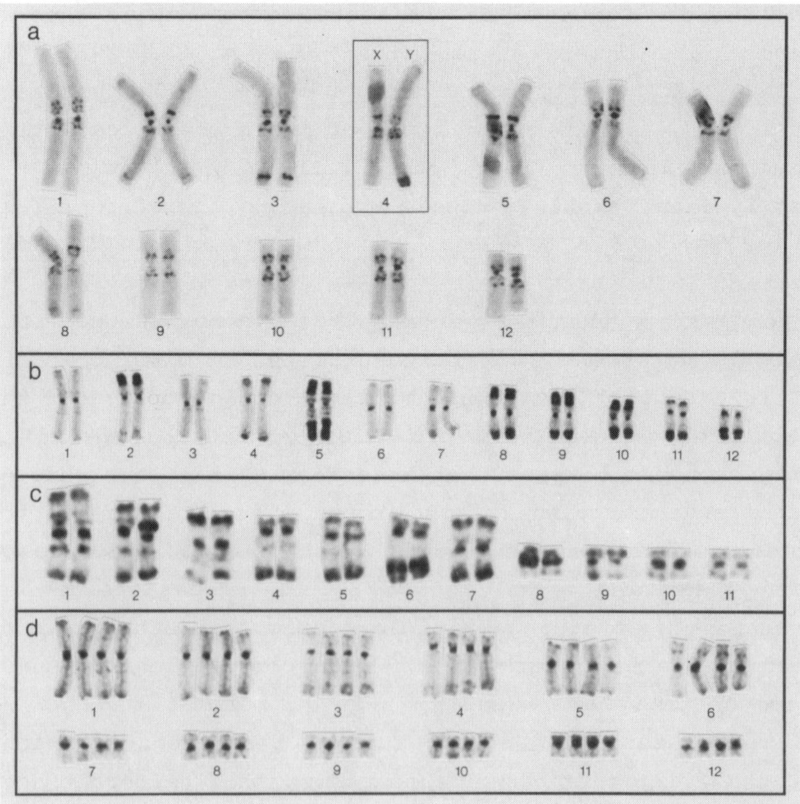

Fig.3. C-banded karyotypes of (a) male *Triturus alpestris*, (b) *Mixophyes fasciolatus*, (c) *Pleurodema brachyops* and (d) the tetraploid *Neobatrachus sudelli*.

always associated with constitutive heterochromatin (Fig. 3). In Urodela and Apoda, the heterochromatic regions are primarily concentrated at centromeric and interstitial chromosome segments (Fig. 3).

Interspecies comparisons of C-band patterns are of basic importance for cytotaxonomic studies. It has been shown that although many species of the various amphibian genera have a shared similarity in chromosome number and morphology, there can be extensive differences in the position and amount of heterochromatin (King, 1980; Schmid, 1978, 1980; Nardi et al., 1986; Sessions & Wiley, 1985). A relationship between the heterochromatic patterns on amphibian chromosomes and evolutionary parameters could not be established, with the exception of the origin of heteromorphic sex chromosomes (see below). It seems possible that the distribution and amount of heterochromatin are variable factors which have significantly contributed to the mechanism of chromosomal isolation, since they change at an extremely rapidly rate in the course of evolution (John, 1988).

A further application of C-banding consists in the analysis of polyploid amphibian karyotypes. Thus, the conventionally stained somatic chromosomes of the tetraploid _Odontophrynus americanus_ can be arranged into sets of four similar chromosomes (quartets); those of the octaploid _Ceratophrys ornata_ into sets of eight similar chromosomes (octets). However, the C-banding patterns revealed heterogeneity within some quartets of _O. americanus_ dividing each of these into two pairs of homologous chromosomes. In analogy, some octets of _C. ornata_ can be subdivided into two quartets of chromosomes with homologous bands (Schmid et al., 1985). These structural heterogeneities within the quartets and octets were interpreted as a "diploidization" of polyploid karyotypes (in the sense of Ohno, 1970). Careful analysis of the conventionally stained tetraploid karyotype of _Xenopus epitropicalis_ shows that the chromosomes can be grouped into sets of four similar chromosomes (quartets). C-banding reveals heterogeneity within _all_ the quartets, dividing each of these into two pairs of homologous chromosomes

(Tymowska & Fischberg, 1982). Finally in the tetraploid species of the genus Neobatrachus, C-banding yields homologous patterns within all quartets (Fig. 3).

The results of the analyses with fluorochromes permit some conclusions about the base pair composition of the repetitive DNA located in the individual heterochromatic C-bands. AT-rich DNA sequences are known to enhance the intensity of quinacrine fluorescence (Weisblum, 1973), whereas mithramycin preferentially forms complexes with GC-rich DNA (Ward et al., 1965). Comparisons among C-banded, quinacrine- and mithramycin-stained karyotypes of diverse species of Anura and Urodela have shown that the heterochromatic regions can be classified into two main types: (1) quinacrine-positive C-bands that are always mithramycin negative ($C^+Q^+MM^-$), and (2) conversely, quinacrine-negative C-bands characterized by bright mithramycin fluorescence ($C^+Q^-MM^+$) (Schmid, 1980; Schmid et al., 1987). This relationship between the two categories of constitutive heterochromatin demonstrates the high AT-content in the C^+Q^+ heterochromatin, and the high GC-content in the C^+MM^+ heterochromatin. The weaker a heterochromatic region fluoresces with quinacrine, the stronger is the intensity of fluorescence achieved with mithramycin, and vice versa (Fig. 2). The sensitivity of these fluorochrome stainings for microscopical detection of AT- and GC-rich DNA in chromosomes is fairly high. Thus, it is estimated to require about 10^7 to $5x10^6$ base pairs (i.e. 0.01 to 0.005 pg DNA) in the case of GC-rich DNA, regardless of the genome size (Schweizer, 1979). Quinacrine-bright heterochromatic regions are very frequently found in amphibian chromosomes (Figs. 1 and 2). This is in contrast to all other vertebrate classes, where C^+Q^+ heterochromatin is rather rare (Jalal et al., 1974; Schmid & Guttenbach, 1988). Considering all the karyotypes banded to date, most of the constitutive heterochromatin in Anura and Urodela is GC-rich. Besides the unequivocal alternative of $C^+Q^+MM^-$ and $C^+Q^-MM^+$ heterochromatin, a third class of C-bands that is Q^- as well as MM^- was found in a careful comparison of banded amphibian chromosomes (Schmid,

1980). It is feasible that alternatively interspersed AT- and GC-rich sequences are present in this special heterochromatin, so that none of the base pair specific fluorochromes yields selective labeling. Heterochromatic regions possessing both Q^+ as well as MM^+ fluorescence have, on the other hand, not yet been discovered in amphibian karyotypes.

NUCLEOLUS ORGANIZER REGIONS

In the karyotypes of all Amphibia, the nucleolus organizer regions (NORs) are the brightest fluorescing regions after staining with GC-specific fluorochromes, such as mithramycin or chromomycin A_3 (Schmid, 1980; Schmid & Guttenbach, 1988). This brilliant fluorescence is caused by the NORs and not by the adjacent NOR-associated heterochromatin. This is best demonstrated by sequential labeling NORs with mithramycin and $AgNO_3$: the bright fluorescence label is always congruent with the dark silver precipitate (Fig. 4). These two staining methods coincide so precisely, that even aberrations and variants of the

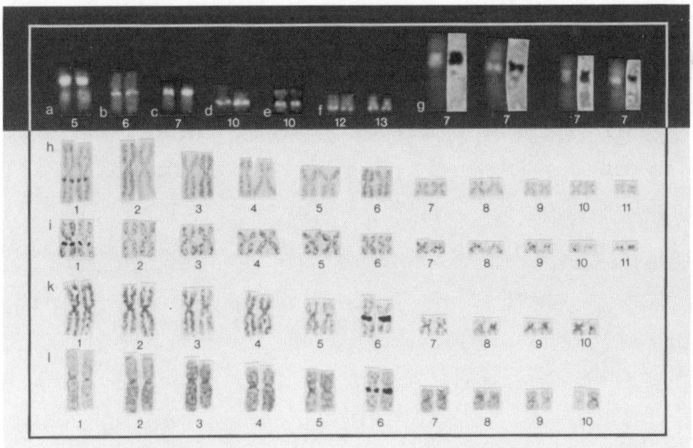

Fig.4. (a-f) Mithramycin-stained NOR-carrying chromosome pairs of (a) *Kaloula pulchra*, (b) *Bufo garmani*, (c) *B. marinus*, (d) *Rana palustris*, (e) *R. macrodon*, and (f) *Chiromantis xerampelina*. (g) Two NOR-carrying chromosome pairs of *B. marinus* stained with mithramycin and subsequently with silver.(h-l) Karyotypes of (h) *Bufo fowleri*, (i) *B. valliceps*, (k) *B. garmani*, and (l) *B. poweri* showing silver-stained NORs.

NORs (duplications, triplications, deletions) can be detected (Schmid, 1980). It is highly unlikely to confuse the NORs with C^+MM^+ heterochromatin, since the latter has a distinctly weaker fluorescence and less defined contrast (Fig. 2).

Positive staining of the 18S+28S ribosomal DNA-containing NORs with GC-specific fluorochromes has also been reported for various species of several plant genera (Schweizer, 1976; Deumling & Greilhuber, 1982), a grasshopper species (Schweizer et al., 1983), and numerous fish, reptiles and birds (Schmid & Guttenbach, 1988). This is not surprising since the 18S+28S rDNA in all eukaryotes tends to have a higher GC content than the main band DNA from the same organism. In addition, the rDNA in all species exhibits high levels of redundancy (Sinclair and Brown, 1971). Therefore, it is interesting to note that only in mammals the NORs do not bind GC-specific fluorochromes, even though the GC content in mammalian rDNA is as high as in other vertebrates (52%-64%; Sinclair and Brown, 1971). Perhaps the total size of the ribosomal DNA cluster within a NOR is the decisive factor in fluorochrome stainability. The fraction of the genome coding for ribosomal RNA in mammals is known to be the smallest among eukaryotes. For example, each of the ten NORs in man contain only 40 to 46 repeating units of ribosomal DNA, whereas *Xenopus laevis* has 400 to 500 of these units located in each of its two NORs (Bross & Krone, 1973; Brown et al., 1972).

Silver (Ag) staining, developed by Goodpasture & Bloom (1975), is an alternative method for examining the number and position of NORs. Due to its specificity and simplicity, Ag-staining has increasingly been employed in cytogenetic studies of vertebrates. In the case of Amphibia, Ag-labeled NORs have been analyzed in a large number of anuran species (King, 1980; Schmid, 1978, 1980; Schmid et al., 1987). It was shown that most of both primitive as well as highly evolved Anura present only one pair of NORs. Interspecies comparisons demonstrated that in the karyotypes of animals belonging to the same or closely related species-groups, the NORs are almost always located at the same chromosome regions (Fig. 4). Exceptions to this rule gave indications of chromosomal rearrangements having occurred in the NOR-carrying chromosome segments during evolution. It is known that the silver staining technique stains neither the rDNA itself, nor the associated rRNA. Cytochemical tests revealed that the Ag-stained material consists of non-histone protein(s) rich in sulfhydril- and disulfide groups which rapidly associate with the newly transcribed rRNA (Howell, 1977; Buys & Osinga, 1980). Residues of this protein associate with the NORs up into the metaphase and can thus be demonstrated with Ag-staining, although at this stage the NORs themselves are transcriptionally inactive (Fan & Penman, 1971). Only those NORs active in the preceding interphase are stained specifically with Ag on mitotic metaphase chromosomes (Miller et al., 1976a,b). Inactive NORs remain undetected by this method. The silver stainability of the NORs reflects their actual transcriptional activity in the interphase nuclei (Hofgärtner et al. 1979; Schmid et al., 1982). In comparison with this, the GC-specific fluorochromes stain the NORs independent of their state of activity.

Considerable differences can be found between the NORs of homologous chromosomes (Fig. 5). This variability is a generalized property of nucleolus organizers and occurs also in

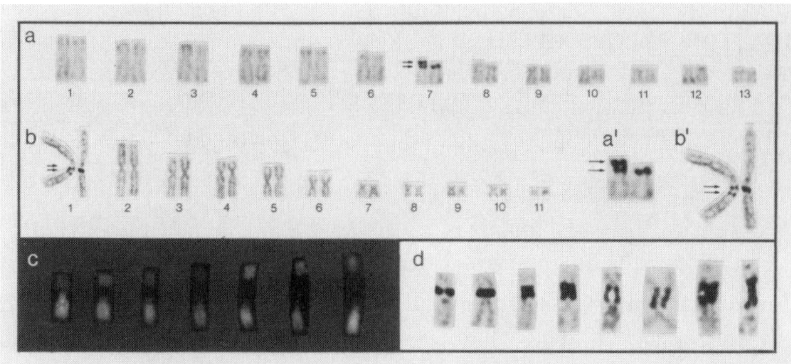

Fig.5. (a,b) Silver-stained karyotypes of (a) *Pelobates syriacus* and (b) *Bufo terrestris* with duplication of one NOR (arrows). (a',b') Enlargements of these chromosomes. (c,d) Variability of NOR-sizes in chromosomes nos. 6 of *Pyxicephalus adspersus*; (c) quinacrine staining, (d) silver staining.

all other vertebrate classes. Unequal meiotic crossovers or sister chromatid exchanges are usually considered to be the cause of the origin of heteromorphic NORs. Furthermore, accidental duplicatons of DNA or a regional disturbance during DNA replication could be responsible for spontaneous additions or deletions of NOR-material (Gold, 1979). In an extensive study on 260 individual animals from 23 genera of Anura, the NORs were examined in detail using Ag-staining and GC-specific fluorochromes (Schmid, 1982). Only 85 specimens (33%) had homologous NORs of equal size. In the remaining 175 animals (67%) the homologous NORs were distinctly heteromorphic. A notable finding was that in 120 (68%) of the 175 animals with unequal NORs, the differences in size could be traced to tandem duplications or triplications affecting one of both NORs. In these cases, two or three Ag-blocks were located on both chromatids (Fig. 5). Furthermore, it was remarkable that among all the animals studied not a single one was found exhibiting a tandem duplication or triplication at both NORs. Apparently, duplicated and triplicated NORs do not occur in homozygous states in wild populations, but rather always are found in combination with a single "normal-sized" NOR on the homologous chromosome (Schmid, 1982).

DNA REPLICATION PATTERNS

5-Bromodeoxyuridine (BrdU) incorporation into chromosomal DNA followed by Hoechst 33258-acridine orange or Giemsa staining (Latt, 1973; Grzeschik et al., 1975) is a refined technique for the analysis of high resolution replication patterns. Although BrdU-replication banding has become one of the most frequently used techniques on mammalian chromosomes, it has been applied only to a limited number of cold-blooded vertebrates. Several possibilities for the use of this technique have been demonstrated on various frog species of the family Ranidae, all differing greatly in their DNA content (Schempp & Schmid, 1981). For example, the homomorphic chromosome pair no. 4 in *Rana es-*

culenta could be identified as sex-specific chromosomes of the XY/XX type. All male animals exhibit an extremely late replicating region in the Y chromosome, which is absent in the females (Fig. 7). In the females, both X chromosomes replicate synchronously. These sex-specific chromosomes cannot be distinguished using other banding techniques. In the highly heteromorphic ZW/ZZ sex chromosomes of Pyxicephalus adspersus, a synchronous replication of the two Z chromosomes of male individuals and a very late replication of the short arm of the W chromosome of females was observed (Fig. 7). These results support the assumption that there is no mechanism of dosage compensation operating for X- or Z-linked genes through sex chromosome inactivation in Amphibia. It was also possible in an urodelan species (Hynobius nigrescens) to induce high-quality BrdU-replication patterns, opening the way for comparative studies within the

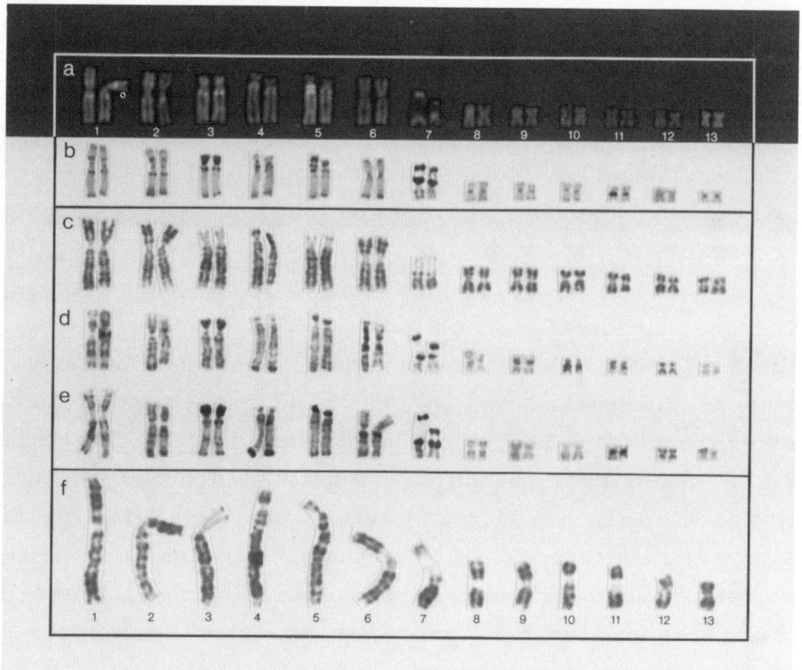

Fig.6. (a-e) Karyotypes of Scaphiobus holbrooki with (a) mithramycin bands, (b) C-bands, (c) early-, (d) mid-, and (e) late BrdU-replication bands. (f) Prometaphase chromosomes with early BrdU-replication bands.

genus Hynobius (Kuro-o et al., 1986).

The differential staining of metaphase chromosomes with Hoechst 33258 and Giemsa after BrdU incorporation during DNA replication results in a pale blue staining of the BrdU-substituted chromosomal segments, whereas thymidine (dT) incorporation produces dark red Giemsa staining. It is a feature of this method that replication patterns on the metaphase chromosomes of a cell can only be demonstrated, if this cell incorporates dT at the beginning of the S-phase, followed by BrdU incorporation into the newly synthesized DNA throughout the remainder of the same S-phase. A brief dT incorporation at the beginning of an S-phase and subsequent BrdU substitution during the longer part of the S-phase manifests so-called early replicating stages. A decrease in the time of BrdU incorporation increases the darkly red stained dT-containing chromosome regions; these banding patterns correspond to the mid-replication stages. Finally, if only the late replicating heterochromatic regions have incorporated BrdU, and are accordingly stained pale blue, so-called late replication stages are revealed (Fig. 6).

With the BrdU-Hoechst-Giemsa technique, multiple replication banding patterns are obtained independent of the DNA content of the species under investigation. Thus, in three species of the family Ranidae having DNA amounts of 13.8 pg (Rana esculenta), 6.8 pg (Rana temporaria), and 4.0 pg (Pyxicephalus adspersus), distinct replication bands were induced even in the very small chromosomes (Schempp & Schmid, 1981). In Figure 6 the early to very late replication bands of Scaphiobus holbrooki are depicted. This anuran has one of the smallest genomes found among vertebrates ($2.1x10^9$ base pairs; Lima-de-Faria, 1983). Nevertheless, very clear replication patterns are resolved along the chromosomes. The detailed analysis of the replication bands shows that the individual bands are identical for the corresponding homologous chromosomes. All the chromosome pairs can therefore be unequivocally identified. The pattern of the replicating regions in the chromosomes of the Amphibia is fundamentally comparable to the sequence from mammalian chromosomes. The C-band-positive regions in the chromosomes of all species examined are very late replicating (Fig. 6).

BANDING OF SEX CHROMOSOMES

Most of the extant Amphibia possess homomorphic sex chromosomes. The number of cytogenetically detected sex chromosomes in the Urodela and Anura is still small when compared with other vertebrates. In the Apoda not a single species with recognizable sex chromosomes has yet been found (for reviews see Schmid, 1983; Schmid and Haaf, 1989). However, improvements of cytogenetic techniques along with high resolution optics have made it possible over the last 10 years to demonstrate the existence of heteromorphic sex chromosomes in initial and advanced stages of morphological differentiation in Amphibia. These studies revealed that both in Anura as well as in Urodela the XY/XX- and ZW/ZZ mechanisms of chromosomal sex determination coexist. Within the same genus, however, the type is uniform. This indicates that during the evolution of Amphibia the XY/XX- or ZW/ZZ system of sex determination was established after the various ancestors of the recent genera had diverged.

The most primitive stage of morphological differentiation between XY chromosomes was found in *Rana esculenta* (see foregoing section and Fig. 7). This was the first example of vertebrate sex chromosomes in which heteromorphism could be recognized ecxlusively by a difference in replication patterns. Several sex chromosomes so far found in Amphibia show that heterochromatinization of the Y- and W chromosomes precedes the actual morphological differentiation. The newts of the genus *Triturus* have 24 meta- to submetacentric chromosomes which can be arranged in 12 homologous pairs after conventional staining. The XY sex chromosomes can only be recognized after specifically staining the constitutive heterochromatin (Schmid et al., 1979; Sims et al., 1984). In the telomeric region of the Y long arm there is always more heterochromatin than in the X long arm (Fig. 3). In the male meiosis of these animals, the long arms of the XY pairs have a highly decreased frequency of chiasmata formation. Similar heteromorphisms between the X and Y were observed in the salamanders from the genus Hydromantes (Nardi et al., 1986) and in the marsupial frog Gastrotheca pseustes (Schmid & Haaf, 1989).

Very highly differentiated XY sex chromosomes are present in the urodelan genera Necturus, Chiropterotriton, Oedipina, Thorius and Bolitoglossa (Sessions, 1980; Sessions & Wiley, 1985; Kezer & Macgregor, 1971). Their Y chromosomes are distinctly

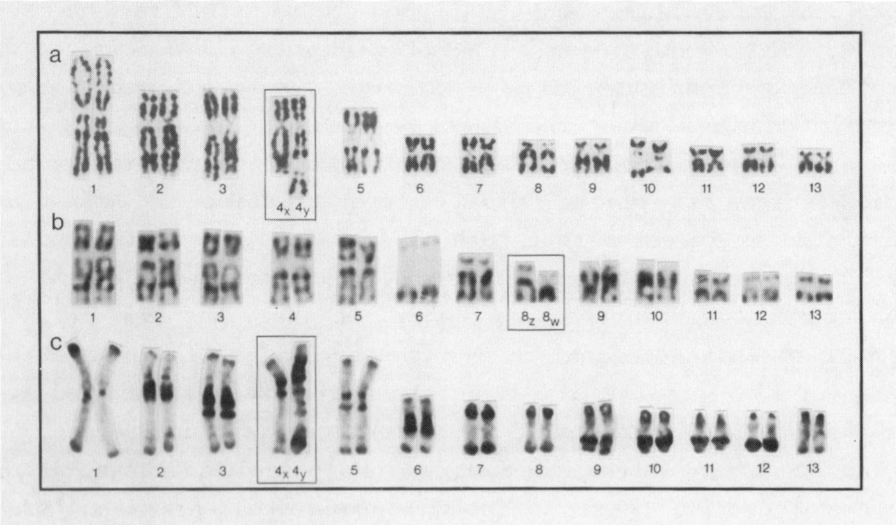

Fig.7. (a,b) Karyotypes of (a) male Rana esculenta and (b) female Pyxicephalus adspersus showing BrdU-replication bands. (c) C-banded karyotype of male Gastrotheca riobambae.

smaller than the X chromosomes and composed almost exclusively of constitutive heterochromatin. The males of the South African bull frog Pyxicephalus delalandii have homomorphic ZZ chromosomes, the females heteromorphic ZW chromosomes. The W is considerably smaller than the Z and its short arm is heterochromatic and very late replicating (Figs. 7 and 8). There is some evidence that the same chromosome pair, which in P. adspersus represents the highly heteromorphic ZW pair, is still in an initial stage of morphologic differentiation in the closely related P. delalandii (Schmid & Haaf, 1989).

Peculiar XY sex chromosomes were found in the marsupial frog <u>Gastrotheca</u> <u>riobambae</u> (Schmid et al., 1983). The Y is the largest element in the karyotype and almost completely heterochromatic (Fig. 7). The only nucleolus organizer is located on the short arm of the X chromosome (Schmid et al., 1986). Since no NORs are located either on the Y chromosome or on any autosome, there is a dosage ratio of 1:2 between male and female animals concerning the 18S+28S ribosomal RNA genes.

The fact that of the sex chromosomes found in Amphibia always the Y- or W chromosome and never the X- or Z chromosome constitutes the derived chromosome (with an increased amount of constitutive heterochromatin) indicates that these went through similar stages in their molecular and morphological differentiation as did the sex chromosomes of snakes, birds and mammals (Singh et al., 1976; 1980).

RESTRICTION ENDONUCLEASE BANDING

In mammalian chromosomes, digestion with certain restriction enzymes that recognize 4 bp or 5 bp long DNA sequences abolishes the Giemsa staining capacity of various chromosomal regions. The type of banding pattern observed by restriction enzyme treatment is specific for each enzyme (for review see Verma & Lubs, 1989). Thus, human chromosomes show simultaneous G-banding and modified C-banding patterns after Hae III or Eco RII digestion, and more typical C-bands following Alu I digestion. The characteristic banding patterns induced were correlated with the specificities of the restriction endonucleases and with the lengths of the DNA fragments which they produce in isolated DNA (Miller et al., 1983; Bianchi et al., 1985).

In Amphibia, the restriction enzymes always induce subsets of the C-banding patterns present in the different karyotypes. The heterochromatic regions can be either resistant or sensitive to the restriction enzyme. The modified C-banding patterns revealed by different restriction endonucleases in the karyotypes of the same species can be either extremely dissimilar or almost completely congruent. Correspondingly, the action of the same restriction enzyme on the karyotypes of different species may vary greatly. Only rarely is there a correlation between the type of C-banding patterns producedby different restriction endonucleases and their specific base pair recognition sequences. In contrast to mammalian and avian chromosomes, restriction enzymes induce no multiple G-banding patterns in amphibian chromosomes. This again was attributed to the difference in DNA organization in the genomes of poikilothermic vertebrates (Schmid & de Almeida, 1988).

IN SITU HYBRIDIZATION WITH SIMPLE REPETITIVE SEQUENCES

Simple repetitive sequences consisting of short tandemly organized motives (e.g. CACACACA or GACAGACAGACAGACA) are ubiquitously distributed throughout eukaryotic genomes. Among many such motives tandem arrays of GATA/GACA simple quadruplets repeats have been demonstrated in every eukaryote investigated so far (Epplen, 1988). The impression of ubiquitous dispersion of simple sequences in a large proportion of amphibian genomes has recently been demonstrated (Schmid et al., 1990). Therefore it seemed possible that such simple sequences are distributed on the chromosome arms in a specific non-random fashion and thus could aid in chromosome indexing. Recently, appyling suitable DNA hybridization probes, banding patterns could be visualized over mammalian metaphase chromosomes without applying banding procedures (Manuelidis and Ward, 1984; Zischler et al., 1989). An orderly arrangement of such sequences in chromosomal DNA would not be apparent readily in the gel analysis of restriction enzyme digested genomic DNA. In order to locate the chromosomal distribution in more detail _in situ_ hybridizations were performed using the $(GACA)_4$ synthetic oligonucleotide as probe. By selecting suitable stringency conditions of hybridization and washing a perfectly matched duplex between oligonucleotides probes and complementary target DNA stretches can be obtained, while mismatches at one or more positions in the duplexes will prevent annealing and signal generation. Thus absolute hybridization specificity can be achieved by synthetic oligonucleotide probes under appropriate conditions. Since radioactively labeled probes need prolonged exposure times and since they are insufficiently precise due to their limited spatial resolution power, a steroid digoxigenin has been attached to the 5'-terminus of the oligonucleotide for _in situ_ hybridization purposes. Subsequently the non-radioactive reporter group is visualized by a polyclonal antibody conjugate directed against the hapten (polyclonal sheep anti-digoxigenin-Fab fragments conjugated to alkaline phosphatase). GATA/GACA quadruplets have been shown to be amplified on the heteromorphic sex chromosomes of many vertebrates (Jones & Singh, 1981). Therefore, the digoxigenin labeled $(GACA)_4$ oligonucleotide was hybridized to the metaphase chromosomes of the South African bull frog _Pyxicephalus adspersus_ to possible distinguish the W chromosome from autosomes. Intriguingly the hybridization sites closely resemble to the C-band pattern to some extent (Fig. 8). In particular the interstitial C-band located on the short arm of chromosome no. 13 hybridizes intensely to $(GACA)_4$.

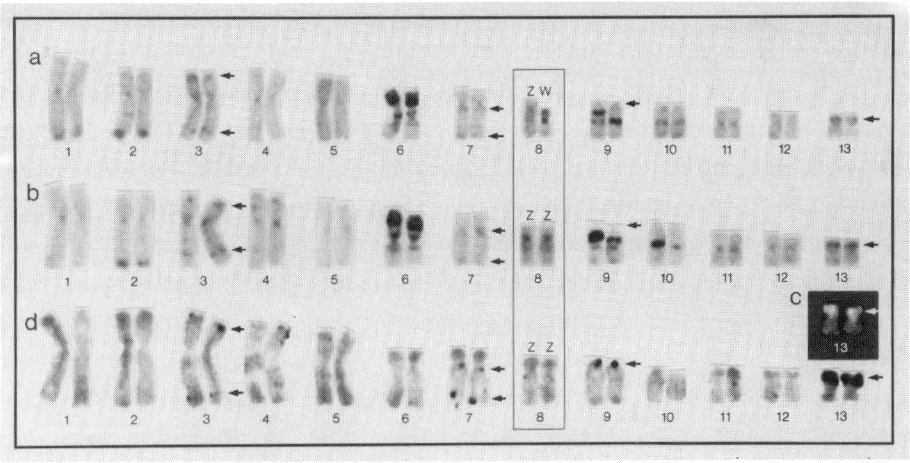

Fig.8. (a,b) Karyotypes of (a) female and (b) male _Pyxicephalus adspersus_. (c) Mithramycin-stained chromosome pair no. 13. (d) Female karyotype after _in situ_ hybridization using a digoxigenin-labeled $(GACA)_4$ probe. The hybridization sites are marked by arrows.

In phase contrast microscopy the hybridization signal seems to encompass the whole short arm of chromosome no. 13 (Fig. 8d). However, careful analysis of many metaphases reveals that GACA sequences are indeed confined to the heterochromatic region. In addition the interstitial C-bands located in the distal arms of chromosome pair no. 3 bears a faint hybridization signal. Further hybridization signals are located on the C-bands of the pericentromeric and long arm telomeric regions of the chromosomes nos. 7. Interestingly, the hybridization sites over the centromeric region of the telocentric chromosomes nos. 9 are the sites which do not stain with Giemsa after alkali treatment indicating that the degree of simple sequence repetition at these loci is not above the threshold level to be detected by the C-banding technique. By contrast, the most prominent blocks of heterochromatin (notably on the chromosomes 6, 9 and W) appear to be devoid of tandem arrays of $(GACA)_n$ simple repeats (Fig. 8).

From the *in situ* hybridization experiments with simple repeats it is apparent that $(GACA)_4$ sequences are arranged largely "en bloc" at discrete loci on the chromosomes in a non-random fashion. The clusters are not equal in intensity which may parallel to the amount of heterochromatin on the particular chromosome region. Lack of the hybridization signal over major heterochromatic blocks reflects that the C-band positive regions are not homogeneous in their molecular organization, though the heterochromatin in *P. adspersus*, irrespective of the chromosomal location, can be labeled specifically by GC-specific fluorochromes (Fig. 8c). Thus, the chromosome-specificity of the subsets of heterochromatin can be demonstrated with synthetic oligonucleotides.

Acknowledgments: Our studies on amphibian chromosomes were supported by the Deutsche Forschungsgemeinschaft (grants Schm 484/2-2 to 2-4, and Schm 484/4-1). M.S. is highly indebted to B. Lumpi for encouragement in preparing the manuscript.

REFERENCES

Bianchi, M.S., Bianchi, N.O., Pantelias,G.E., and Wolff,S (1985) The mechanism and pattern of banding induced by restriction endonucleases in human chromosomes. Chromosoma 91: 131-136.

Bross, K., and Krone, W. (1973) Ribosomal cistrons and acrocentric chromosomes in man. Hum. Genet. 18: 71-75.

Brown, D.D., Wensink, P.C., and Jordan, E. (1972) A comparison of the ribosomal DNA's of Xenopus mulleri: the evolution of tandem genes. J. Mol. Biol. 63: 57-73.

Buys, C.H.C.M., and Osinga, J. (1980) Abundance of protein--bound sulfhydril and disulfide groups at chromosomal nucleolus organizing regions. A cytochemical study on the selective silver staining of NORs. Chromosoma 77: 1-11.

Comings, D.E. (1978) Mechanism of chromosome banding and implications for chromosome structure. Ann. Rev. Genet. 12: 25-46.

Comings, D.E., and Drets, M. E. (1976) Mechanisms of chromosome banding. IX. Are variations in DNA base composition adequate to account for quinacrine, Hoechst 33258, and daunomycin banding? Chromosoma 52: 229-243.

Cuny, G., Soriano, P., Macaya, G., and Bernardi, G. (1981) The major components of the mouse and human genomes: preparation, basic properties, and compositional heterogeneity. Eur. J. Biochem. 111: 227-233.

Deumling, B., and Greilhuber, J. (1982) Characterization of heterochromatin in different species of the Scilla siberia group (Liliaceae) by in situ hybridization of satellite DNA and fluorochrome banding. Chromosoma 84: 535-555.

Epplen, J.T (1988) On simple repeated GATA/GACA sequences: a critical reappraisal. J. Hered. 79: 409-417.

Fan, H., and Penman, S. (1971) Regulation of synthesis and processing of nucleolar components in metaphase-arrested cells. J. Mol. Biol. 59: 27-42.

Filipski, J., Thiery, J.P., and Bernardi, G. (1973) An analysis of the bovine genome by Cs_2SO_4/Ag^+ density gradient centrifugation. J. Mol. Biol. 80: 177-197.

Gold, J.R. (1979) Cytogenetics. In: Fish Physiology, vol. 8. Academic Press, New York.

Goodpasture, C., and Bloom, S.E. (1975) Visualization of nucleolar organizer regions in mammalian chromosomes using silver staining. Chromosoma 53: 37-50.

Grzeschik, K-H., Kim, M.A., and Johansmann, R. (1975) Late replication bands of human chromosomes demonstrated by fluorochrome and Giemsa staining. Hum. Genet. 29: 41-59.

Hofgärtner, F.J., Schmid, M., Krone, W., Zenzes, M.T., and Engel, W. (1979) Pattern of activity of nucleolus organizers during spermatogenesis in mammals as analyzed by silver-staining. Chromosoma 71: 197-216.

Holmquist, G. (1988) DNA sequences in G-bands and R-bands. In: Chromosome and Chromatin Structure (K.W. Adolph, Ed), CRC Press, Boca Raton, Florida.

Howell, W.M. (1977) Visualization of ribosomal gene activity: silver stains proteins associated with rRNA transcribed

from oocyte chromosomes. Chromosoma 62: 361-367.
Hudson, A.P., Cuny, G., Cortadas, J., Haschemeyer, A.E.V., and Bernardi, G. (1980) An analysis of fish genomes by density gradient centrifugation. Eur. J. Biochem. 112: 203-210.
Jalal, S.M., Clark, R.W., Hsu, T.C., and Pathak, S. (1974) Cytological differentiation of constitutive heterochromatin. Chromosoma 48: 391-403.
John, B. (1988) The biology of heterochromatin. In: Heterochromatin. Molecular and Structural Aspects (R.S. Verma, Ed), Cambridge University Press, Cambridge, pp. 1-147.
Jones, K.W., and Singh, L. (1981) Conserved repeated DNA sequences in vertebrate sex chromosomes. Hum. Genet. 58: 46 – 53.
Kezer, J., and Macgregor, H.C. (1971) A fresh look at meiosis and centromeric heterochromatin in the red-backed salamander Plethodon c. cinereus (Green). Chromosoma 33: 146-166.
King, M. (1980) C-banding studies on Australian hylid frogs: secondary constriction structure and the concept of euchromatin transformation. Chromosoma 80: 191-217.
Kuro-o, M., Ikebe, C., and Kohno, S (1986) Cytogenetic studies on Hynobiidae (Urodela). IV. DNA replication bands (R-banding) in the genus Hynobius and the banding karyotype of Hynobius nigrescens Stejneger. Cytogenet. Cell Genet. 43: 14-18.
Latt, S.A. (1973) Microfluorometric detection of deoxyribonucleic acid replication in human metaphase chromosomes. Proc. Natl. Acad. Sci. USA 70: 3395-3399.
Latt, S.A. (1977) Fluorescent probes of chromosome structure and replication. Can. J. Genet. Cytol. 19: 603-623.
Latt, S.A., Munroe, S.H., Disteche, C., Rogers, W.E., and Cassel, D.M. (1977) Uses of fluorescent dyes to study chromosome structure and replication. In: Chromosomes Today, vol. 6. (A. de la Chapelle and M. Sorsa, Eds), Oxford Chromosome Conference, Edinburgh, pp. 27-36.
Lima-de-Faria, A. (1983) Molecular Evolution and Organization of the Chromosome. Elsevier, Amsterdam, pp. 275-294.
Macaya, G., Thiery, J.P., and Bernardi, G. (1976) An approach to the organization of the eukaryotic genomes at a macromolecular level. J. Mol. Biol. 108: 237-254.
Macgregor, H.C., and Varley, J.M. (1988) Working with Animal Chromosomes. John Wiley, Chichester.
Manuelidis, L., and Ward, D.C. (1984) Chromosomal and nuclear distribution of the Hind III 1.9-kb human DNA repeat segment. Chromosoma 91: 28-38.
Medrano, L., Bernardi, G., Couturier, J., Dutrillaux, B., and Bernardi, G. (1988) Chromosome banding and genome compartmentalization in fishes. Chromosoma 96: 178-183.
Miller, D.A., Choi, Y.C., and Miller O.J. (1983) Chromosome localization of highly repetitive human DNA's and amplified ribosomal DNA with restriction enzymes. Science 219: 395-397.
Miller, D.A., Dev, V.G., Tantravahi, R., and Miller, O.J. (1976a) Suppression of human nucleolus organizer activity in mouse-human somatic hybrid cells. Exp. Cell Res. 101: 235 243.

Miller, O.J., Miller, D.A., Dev, V.G., Tantravahi, R., and Croce, C.M. (1976b). Expression of human and suppression of mouse nucleolus organizer activity in mouse-human somatic cell hybrids. Proc. Natl. Acad. Sci. USA 73: 4531-4535.

Nardi, I., Andronico, F., De Lucchini, S., and Batistoni, R. (1986) Cytogenetics of the European plethodontid salamanders of the genus Hydromantes (Amphibia, Urodela). Chromosoma 94: 337-388.

Ohno, S. (1970) Evolution by Gene Duplication. Springer, Berlin.

Olofsson, B., and Bernardi, G. (1983) Organization of nucleotide sequences in the chicken genome. Eur. J. Biochem. 130: 241-245.

Pizon, V., Cuny, G., and Bernardi, G. (1984) Nucleotide sequence organization in the very small genome of a tetraodontid fish, Arothron diadematus. Eur. J. Bioch. 140: 25-30.

Rooney, D.E., and Czepulkowski, B.H. (1986) Human Cytogenetics. A Practical Approach. IRL Press, Oxford.

Schempp, W., and Schmid, M. (1981) Chromosome banding in Amphibia. VI. BrdU-replication patterns in Anura and demonstration of XX/XY sex chromosomes in Rana esculenta. Chromosoma 83: 697-710.

Schmid, M. (1978) Chromosome banding in Amphibia. I. Constitutive heterochromatin and nucleolus organizer regions in Bufo and Hyla. Chromosoma 66: 361-388.

Schmid, M. (1980) Chromosome banding in Amphibia. IV. Differentiation of GC- and AT-rich chromosome regions in Anura. Chromosoma 77: 83-103.

Schmid, M. (1982) Chromosome banding in Amphibia. VII. Analysis of the structure and variability of NORs in Anura. Chromosoma 87: 327-344.

Schmid, M. (1983) Evolution of sex chromosomes and heterogametic systems in Amphibia. Differentiation 23 (Supplement): S13-S22.

Schmid, M., and de Almeida, C.G. (1988) Chromosome banding in Amphibia. XII. Restriction endonuclease banding. Chromosoma 96: 283-290.

Schmid, M., and Guttenbach, M. (1988) Evolutionary diversity of reverse (R) fluorescent chromosome bands in vertebrates. Chromosoma 97: 101-114.

Schmid, M., and Haaf, T. (1989) Origin and evolution of sex chromosomes in Amphibia: the cytogenetic data. In: Evolutionary Mechanisms in Sex Determination (S.S. Wachtel, Ed), CRC Press, Boca Raton, Florida, pp. 37-56.

Schmid, M., Haaf, T., Geile, B., and Sims, S. (1983) Chromosome banding in Amphibia. VIII. An unusual XY/XX-sex chromosome system in Gastrotheca riobambae (Anura, Hylidae). Chromosoma 88: 69-82.

Schmid, M., Haaf, T., and Schempp, W. (1985) Chromosome banding in Amphibia. IX. The polyploid karyotypes of Odontophrynus americanus and Ceratophrys ornata (Anura, Leptodactylidae).

Schmid, M., Löser, C., Schmidtke, J., and Engel, W. (1982) Evolutionary conservation of a common pattern of activity of nucleolus organizers during spermatogenesis. Chromosoma 86: 149-179.

Schmid, M., Nanda, I., Steinlein, C., Kausch, K., and Epplen, J.T. (1990) Sex determining mechanisms and sex chromosomes in Amphibia. In: Amphibian Cytogenetics and Evolution (S.K. Sessions and D. Green, Eds), Academic Press, New York, in press.

Schmid, M., Olert, J., and Klett, C. (1979) Chromosome banding in Amphibia. III. Sex chromosomes in Triturus. Chromosoma 71: 29-55.

Schmid, M., Sims, S.H., Haaf, T., and Macgregor, H.C. (1986) Chromosome banding in Amphibia. X. 18S and 28S ribosomal RNA genes, nucleolus organizers and nucleoli in Gastrotheca riobambae. Chromosoma 94: 139-145.

Schmid, M., Vitelli, L., and Batistoni, R. (1987) Chromosome banding in Amphibia. XI. Constitutive heterochromatin, nucleolus organizers, 18S+28S and 5S ribosomal RNA genes in Ascaphidae, Pipidae, Discoglossidae and Pelobatidae. Chromosoma 95: 271-284.

Schweizer, D. (1976) Reverse fluorescent chromosome banding with chromomycin and DAPI. Chromosoma 58: 307-324.

Schweizer, D. (1979) Fluorescent chromosome banding in plants: applications, mechanisms, and implications for chromosome structure. In: Proc. 4th John Innes Symp.: The Plant Genome (D.R. Davies and R.A. Hopwood, Eds), John Innes Charity, Norwich, pp. 61-72.

Schweizer, D., Loidl, J., and Hamilton, B. (1987) Heterochromatin and the phenomenon of chromosome banding. In: Results and Problems in Cell Differentiation, vol. 14, Structure and Function of Eukaryotic Chromosomes (W. Hennig, Ed), Springer, Berlin, pp. 235-254.

Schweizer, D., Mendelak, M., White, M.J.D., and Contreras, N. (1983) Cytogenetics of the parthenogenetic grasshopper Warramaba virgo and its bisexual relatives. X. Patterns of fluorescent banding. Chromosoma 88: 227-236.

Sessions, S.K. (1980) Evidence for a highly differentiated sex chromosome heteromorphism in the salamander Necturus maculosus. Chromosoma 77: 157-168.

Sessions, S.K., and Wiley, J.E. (1985) Chromosome evolution in salamanders of the genus Necturus. Brimleyana 10: 37-52.

Sims, S.H., Macgregor, H.C., Pellat, P.S., and Horner, H.A. (1984) Chromosome 1 in crested and marbled newts (Triturus): an extraordinary case of heteromorphism and independent chromosome evolution. Chromosoma 89: 169-185.

Sinclair, J.H., and Brown, D.D. (1971) Retention of common nucleotide sequences in the ribosomal deoxyribonucleic acid of eukaryotes and some of their physical characteristics. Biochemistry 10: 2761-2769.

Singh, L., Purdom, I.F., and Jones, K.W. (1976) Satellite DNA and evolution of sex chromosomes. Chromosoma 59: 43-62.

Singh, L., Purdom, I.F., and Jones, K.W. (1980) Sex chromosome associated DNA: evolution and conservation. Chromosoma 79: 137-157.

Thiery, J.P., Macaya, G., and Bernardi, G. (1976) An analysis of eukaryotic genomes by density gradient centrifugation. J. Mol. Biol. 108: 219-235.

Tymowska, J., and Fischberg, M. (1982) A comparison of the karyotype, constitutive heterochromatin, and nucleolar organizer regions of a new tetraploid species Xenopus epitropicalis Fischberg and Picard with those of Xenopus tropicalis Gray (Anura, Pipidae). Cytogenet. Cell Genet. 34: 149-157.

Verma, R.S., and Babu, A. (1989) Human Chromosomes. Manual of Basic Techniques. Pergamon Press, New York.

Ward, D.C., Reich, E., and Goldberg, I.H. (1965) Base specificity in the interaction of polynucleotides with antibiotic drugs. Science 149: 1259-1263.

Weisblum, B. (1973) Fluorescent probes of chromosomal DNA structure: three classes of acridines. Cold Spring Harbor Symp. Quant. Biol. 38: 441-449.

Zischler, H., Nanda, I., Schäfer, R., Schmid, M., and Epplen, J.T. (1989) Digoxigenated oligonucleotide probe specific for simple repeats in DNA fingerprinting and hybridization in situ. Hum. Genet. 82: 227-233.

Cytogenetics of Amphibians and Reptiles
Advances in Life Sciences

BANDING KARYOTYPES AND PHYLOGENETIC RELATIONSHIPS OF HYNOBIIDAE FROM JAPAN AND KOREA

S. Kohno[1], C. Ikebe[2] and M. Kuro-o[1]

1: Department of Biology, Faculty of Science, Toho University; 2: Department of Biology, Faculty of General Education, Toho University, Funabashi, Chiba 274, Japan.

SUMMARY: Banding analyses were performed on Japanese and Korean hynobiid salamanders, and multiple bands were observed in the arms of chromosomes from 14 species in two genera, *Hynobius* and *Salamandrella*, using C-banding (CBG) and R-banding (RBG) techniques. Identification was made on the basis of banding patterns of 18 out of 28 chromosome pairs (2n=56) in 10 species of *Hynobius* (*H. leechii*, *H. nebulosus*, *H. tokyoensis*, *H. nigrescens*, *H. tsuensis*, *H. dunni*, *H. abei*, *H. takedai*, *H. lichenatus*, and *H. okiensis*), 18 out of 29 chromosome pairs (2n=58) of *H. boulengeri* and *H. kimurae*, 16 out of 20 chromosome pairs (2n=40) in *H. retardatus*, and 20 out of 31 chromosome pairs (2n=62) of *Salamandrella keyserlingii*. The banding patterns of the karyotypes in 12 *Hynobius* species having 56 and 58 chromosomes were quite similar, except for the terminal bands of chromosome 2 and short arms of chromosome 10. Among these 12 species, about 90% of total chromosome length of the karyotypes in each species showed homoeologous banding patterns. Among 13 species including *H. retardatus*, homoeology decreased to about 65% (14 homoeologous chromosome pairs including 3 that were partially homoeologous). In all 14 species including *S. keyserlingii*, homoeology further decreased to about 15% (only 4 homoeologous chromosome pairs including 2 partially homoeologous). Still, the presence of the homoeologous chromosomes between these two genera, *Hynobius* and *Salamandrella*, is karyological evidence that the species have differentiated from a common ancestor.

Salamanders belonging to the family Hynobiidae have wide distribution, extending from eastern Europe to eastern Asia. This family may possibly be the most primitive of living Urodeles (Morescalchi, 1973; Duellman and Trueb, 1986), and includes the genera, Hynobius, Ranodon, Batrachuperus, Salamandrella, Liua, Onychodactylus, Pachyhynobius and Paradactylodon (Brame, 1985).

Salamanders belonging to the genus Hynobius are distributed in eastern Asia. To date, one and fifteen species belonging to the genus Hynobius from the Korean peninsula and Japan have been described (Sato, 1943; Matsui and Miyazaki, 1984; Ikebe et al., 1986; Matsui, 1987).

Chromosome analyses of the genus Hynobius have been rather extensively carried out primarily by the conventional Giemsa method (Azumi and Sasaki, 1971; Morescalchi et al., 1979; Ikebe and Kohno, 1979a, 1979b; Seto et al., 1983, 1988; Seto and Matsui, 1984; Matsui et al., 1985; Ikebe et al., 1986). The results indicate the chromosome numbers to be 2n=56 and 58 in most Hynobius, and karyotypes to be essentially the same, though minor differences do exist. On the one hand, chromosome number is 2n=40 in H. retardatus, the chromosome constitution of which is considerably different from that of other species in Hynobius (Azumi and Sasaki, 1971).

The salamander distributed widely from eastern Europe to eastern Asia, Salamandrella keyserlingii, and first classified by Dybowski (1870) was reclassified into the genus Hynobius on the basis of adult external characters (Dunn, 1923; Thorn, 1968). However, classification was made yet again on the basis of skull features so that, at the present, it belongs to the genus Salamandrella (Sato, 1943; Nakamura and Ueno, 1963). The chromosome number of this species was found to be 62 as a diploid set (Morescalchi et al., 1975, 1979; Wang et al.; 1983) and Morescalchi et al. (1979) observed the karyotype of Salamandrella keyserlingii to be similar to those of the genera Ranodon and Batrachuperus, but to differ from that of the genus, Hynobius.

Recently, Kohno et al. (1983) and Kuro-o et al (1986) were able to obtain multiple bands in Hynobius species by C-banding and modified R-banding treatment. By these techniques, a number of karyotypes of Hynobius species was analyzed (Kohno et al., 1986; Seto et al., 1986; Kuro-o et al., 1987a, 1987b; Ikebe et al., 1987; Izumisawa et al., in press). However, additional karyotype analysis of the genus Hynobius and related species should be carried out.

This paper presents C-banding and the modified R-banding techniques improved for the study of genus Hynobius. A comparison of the banding patterns in thirteen Hynobius and one Salamandrella was made and the phylogenetic relationships among them are discussed.

The Thesis of Master of Science, Toho University and its appendix written by Kuro-o, M. (1986) has contributed most significantly to the present study.

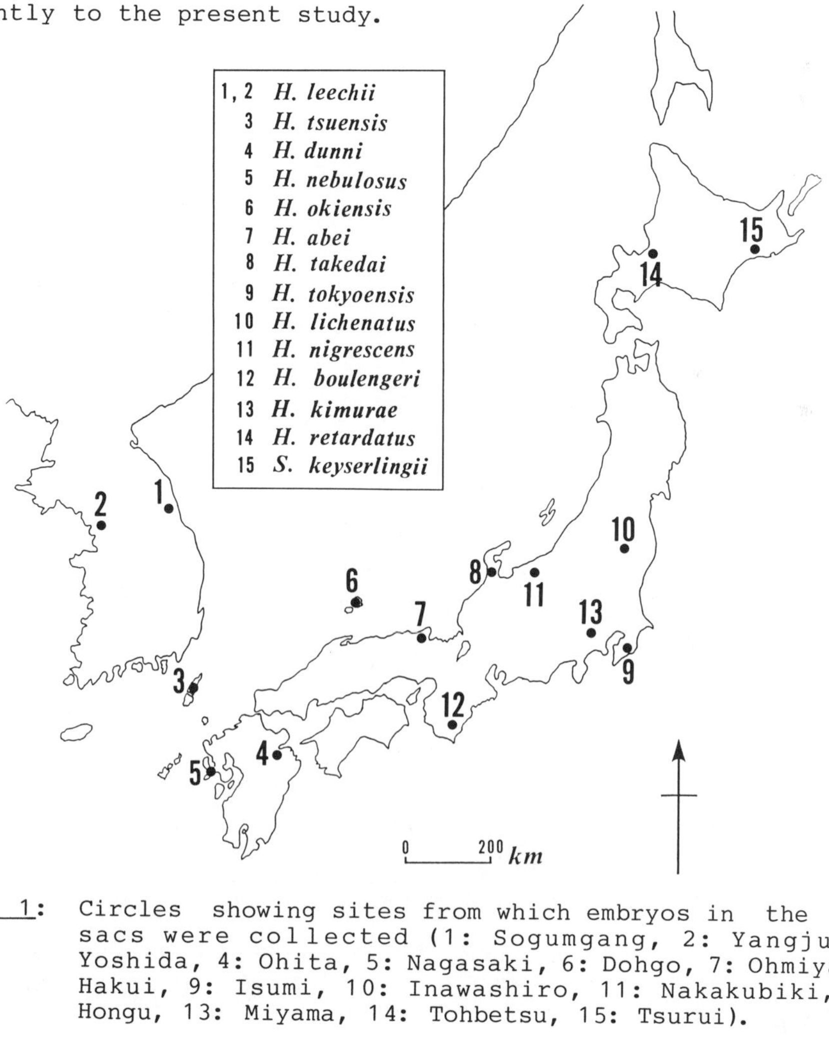

Fig. 1: Circles showing sites from which embryos in the egg-sacs were collected (1: Sogumgang, 2: Yangju, 3: Yoshida, 4: Ohita, 5: Nagasaki, 6: Dohgo, 7: Ohmiya, 8: Hakui, 9: Isumi, 10: Inawashiro, 11: Nakakubiki, 12: Hongu, 13: Miyama, 14: Tohbetsu, 15: Tsurui).

MATERIALS AND METHODS

Materials: Embryos of *H. leechii*, *H. tsuensis*, *H. dunni*, *H. nebulosus*, *H. okiensis*, *H. abei*, *H. takedai*, *H. tokyoensis*, *H. lichenatus*, *H. nigrescens*, *H. boulengeri*, *H. kimurae*, *H. retardatus* and *S. keyserlingii* in egg-sacs were collected from Sogumgang and Yangju in Korea: *H. le.*, Yoshida (Nagasaki): *H. ts.*, Ohita (Ohita): *H. d.*, Nagasaki (Nagasaki): *H. ne.*, Dohgo (Shimane): *H. o.*, Ohmiya (Kyoto): *H. a.*, Hakui (Ishikawa): *H. ta.*, Isumi (Chiba): *H. to.*, Inawashiro (Fukushima): *H. li.*, Nakakubiki (Niigata): *H. ni.*, Hongu (Wakayama): *H. b.*, Miyama (Tokyo): *H. k.*, Tohbetsu (Hokkaido): *H. r.*, Tsurui (Hokkaido): *S. k.* in Japan, (Fig. 1).

Preparation of chromosomes and C-banding (CBG) treatment: Fertilized eggs were incubated at our laboratory. Embryos were removed with care from the chorion with tweezers and scissors, and kept alive for 24 hrs at 20° C in a 0.5% colchicine solution. The embryos were subsequently thorough disrupted with Pasteur pipettes and their cells were treated with hypotonic solution (amphibian saline : distilled water = 1:15) for 30-45 minutes, followed by fixation (methanol : acetic acid = 3:1) and air drying on glass slides.

C-banding patterns were obtained according to the method of Sumner (1971) with slight modification as follows. One or 2 days following preparation, the slides were dipped in a solution of 0.2N hydrochloric acid (20° C) for 30-50 minutes and rinsed in distilled water. This was followed by incubation in 5% barium hydroxide at 50° C for 10 minutes and rinsing in distilled water. Incubation was conducted for an additional hour in 0.3 M sodium chloride-0.03 M sodium citrate (2 x SSC) at 60° C. After being rinsed in distilled water, the slides were stained with 4% Giemsa solution in phosphate buffered saline at pH 6.8 for 30-50 minutes, rinsed again in distilled water and dried at room temperature.

R-banding (RBG) treatment: The R-banding method used here was that by Dutrillaux et al. (1973) with some modification (Kuro-o et al., 1986). The embryos were incubated for a few days at 20° C before culture. One embryo was carefully removed from the chorion with sterilized needles, washed several times in sterile amphibian saline, and transferred to a culture dish (35 mm in diameter, 10 mm in height) containing 1.5 ml of medium comprised of 60% Eagle's MEM, 20% calf serum, 20% double-distilled water, and 400 μg/ml 5-bromodeoxyuridine (BrdU). The embryos were thoroughly disrupted with a Pasteur pipette and their cells were incubated for 33 hours at 20° C in a dark and humidified incubator into which air, containing 2% CO_2 was introduced at a constant rate. Eight hours before completion of incubation, 1.5 ml of the above-mentioned medium containing 1% colchicine were added so that its final concentration was 0.5%. After incubation, the cells were treated with hypotonic solution (amphibian saline : distilled water = 1:7) for 55 minutes at room temperature, followed by fixation (methanol : acetic acid = 3:1) and drying on glass slides with a flame.

The slides were stained by FPG (fluorescence plus Giemsa) technique (Perry and Wolff, 1974) modified slightly as follows. The slides were dried for a few days at room temperature and stained with 50 μg/ml 33258 Hoechst in Ca^{2+}- and Mg^{2+}-free phosphate buffered saline (PBS-) for 15 minutes, followed by brief rinsing in distilled water, mounting with PBS-, and exposure to black-light-blue light at a distance of 10 cm for 30 minutes. The cover slips were removed and the slides were rinsed briefly in water, and incubated in 2 x SSC for 30 minutes at 60° C. They were then thoroughly rinsed in running water and stained with 3% Giemsa solution in phosphate buffered saline at pH 6.8 for 4 minutes.

RESULTS

By C-banding and modified R-banding, it was possible to discern multiple clear C-bands and R-bands on the arms of chromosomes in more than 20 metaphases of each species. Examples of banded

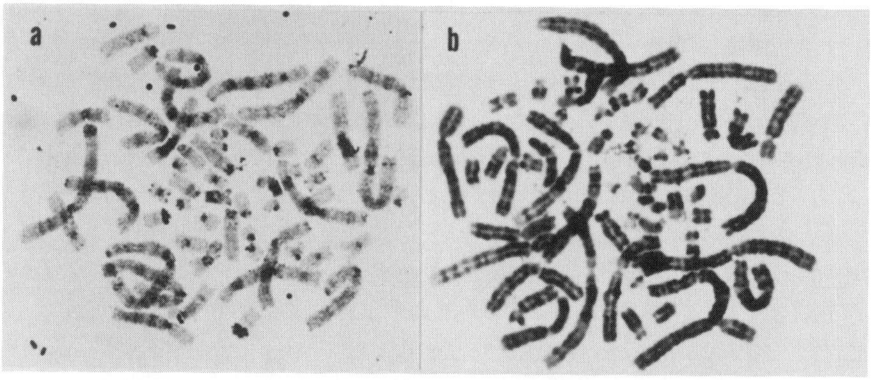

Fig. 2: C-banded metaphase from *Hynobius tsuensis* (a) and R-banded metaphase from *H. retardatus* (b).

metaphases for each banding method are shown in Fig. 2.

These banding patterns made possible the identification of 18 out of 28 or 29 chromosome pairs of 12 *Hynobius* species except for *H. retardatus*. Nine of these are large-size (chromosomes 1-9), 4 medium-size (Chromosomes 10-13), and 5 out of 15 small-size (chromosomes 14, 15, and three larger acrocentric chromosome pairs, 19-21 or 20-22). The remaining 10 or 11 small-sized chromosome pairs showed no definite banding patterns. In the case of *H. retardatus*, 16 out of 20 chromosome pairs could be identified (large-size: 1-10, medium-size: 11-13, small-size: 14-16), while no identification could be made of the remaining 4 pairs. In *S. keyserlingii*, 20 out of 31 chromosome pairs were easily identified: 2 metacentric, 1 submetacentric, 4 subtelocentric, and 13 acrocentric chromosome pairs, according to the quantitative definition of Levan *et al.* (1964). The remaining 11 chromosome pairs failed to show definit banding patterns. Examples of banded karyotypes of *H. nebulosus*, *H. kimurae*, *H. retardatus* and *S. keyserlingii* are shown in Figs. 3(a-d).

Chromosome pairs identified by banding patterns are shown in Table I along with their identification numbers by which whole or partial chromosomal homoeology is indicated. Among the 13 *Hynobius* species, 11 chromosome pairs showed respective identical banding patterns: chromosomes 1, 3, 6, 7, 8, 9, 11, 13, 14(except for that in *H. kimurae*), 15 and 21(or 20) were homoeologous to

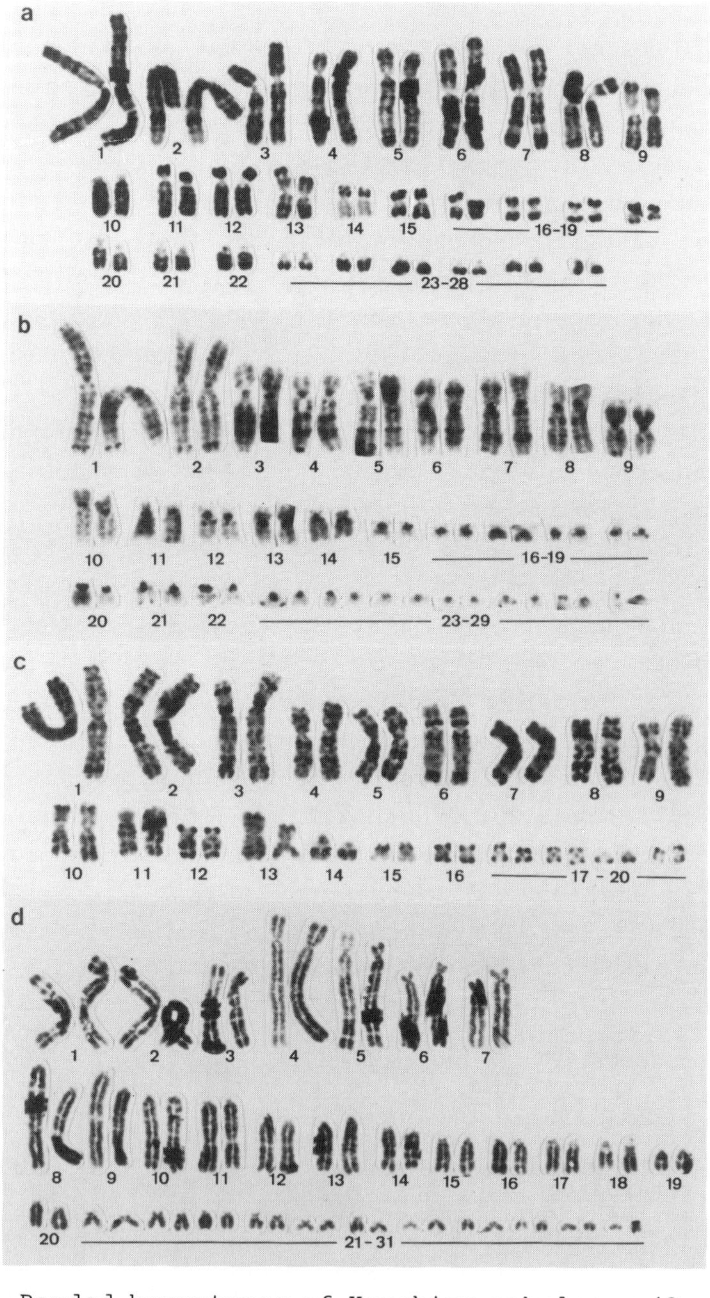

Fig. 3: Banded karyotypes of *Hynobius nebulosus* (2n=56):a (R-banding), *Hynobius kimurae* (2n=58): b (C-banding), *Hynobius retardatus* (2n=40): c (R-banding) and *Salamandrella keyserlingii* (2n=62): d (R-banding).

chromosomes 1, 4, 5, 7, 9, 10, 12, 13, 15, 16, and 14 in *H. retardatus*, respectively. Three chromosome pairs were partially homoeologous; in Fig. 4, the regions along the vertical bars on chromosomes 4 and 5, and long arms of chromosomes 10 in the 12 *Hynobius* species correspond to those on chromosomes 6 and 3, and the long arm of chromosome 11 in *H. retardatus*, respectively. Whole short arms of chromosomes 10 in the 12 species and that of chromosome 11 in *H. retardatus* were stained C-positive and R-negative, and showed variation in length. This variation is discussed by Ikebe *et al.* (1990) in relation to phylogenetic relationships in the genus, *Hynobius*.

In addition to the above findings, three chromosome pairs showed respective identical banding patterns (chromosomes 12, 20 and 22, or 12, 19 and 20) among the 12 species. The identical banding patterns of these three chromosome pairs could not be found in *H. retardatus*. Length was clearly found to differ in C-positive and R-negative bands at the terminals of the long arms on chromosome 2 in the 12 species.

In Fig. 4 and Table I, chromosomes 2, 17, 18 and 19 in *S. keyserlingii* are homoeologous to chromosome 7 and three small acrocentric chromosome pairs in the 12 *Hynobius* species (ex. *H. leechii*: chromosome 20, 21 and 22). Chromosomes 3 and 5 in *S. keyserlingii* were partially homoeologous to chromosome 9 in the 12 *Hynobius* species and chromosome 10 in *H. retardatus*, and chromosome 3 in the 12 *Hynobius* species and chromosome 4 in *H. retardatus*, respectively.

Fig. 4 (facing page): Comparison of R-banding patterns in 6 species, *Hynobius leechii*, *H. nebulosus*, *H. tokyoensis*, *H. nigrescens*, *H. retardatus* and *Salamandrella keyserlingii*. The lower case letters under the chromosomes indicate species (l: *H. leechii*, nb: *H. nebulosus*, t: *H. tokyoensis*, ng: *H. nigrescens*, r: *H. retardatus*, k: *S. keyserlingii*). Numbers are for chromosome identification in each species. Numbers in the left superscript position identify chromosomes common to 4 species, *H. leechii*, *H. nebulosus*, *H. tokyoensis*, and *H. nigrescens*. When the numbers correspond to 5 *Hynobius* species, none are indicated for *H. retardatus* (numbers on right side). The vertical bars indicate identical regions in replication patterns. Dashed lines are drawn through the centromeres.

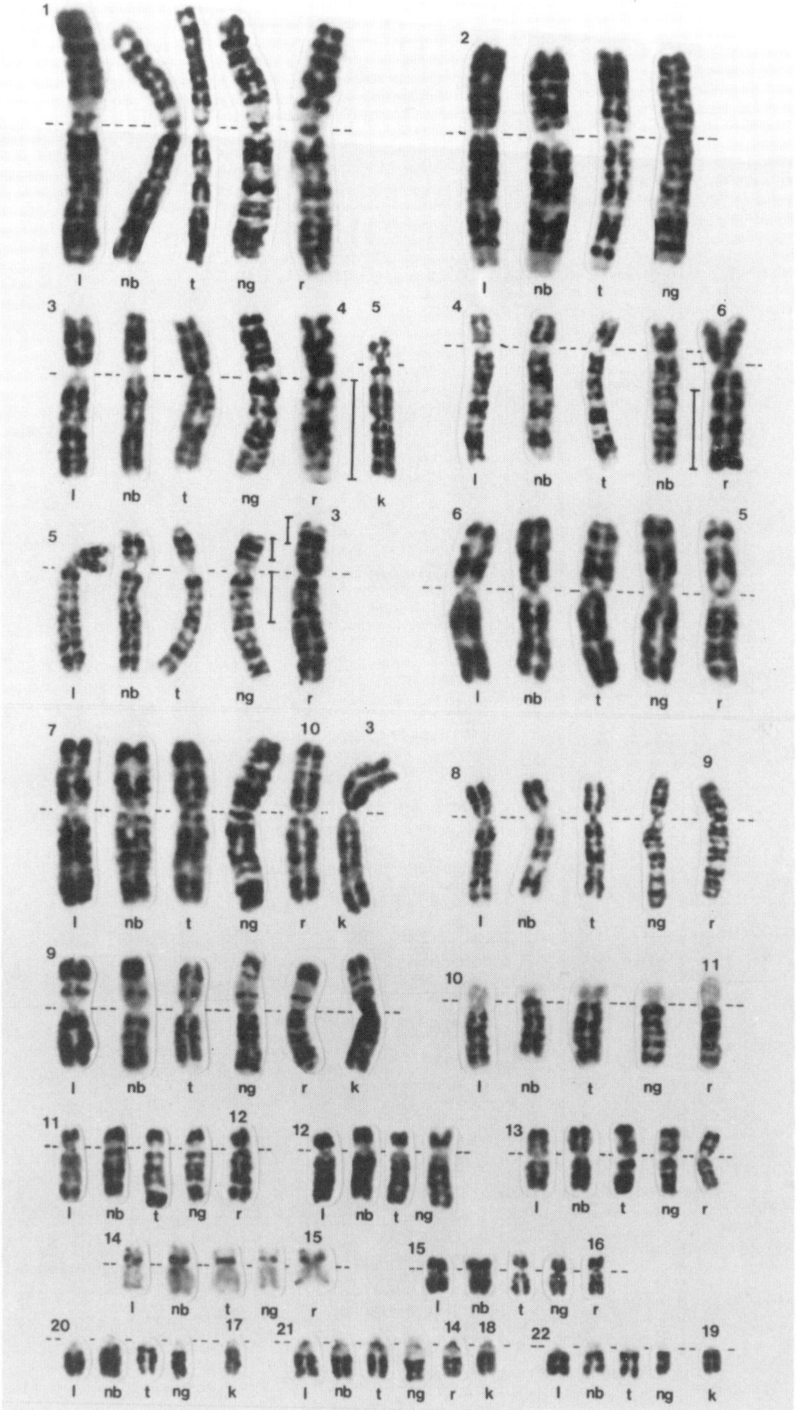
1
l nb t ng r
2
l nb t ng
3
l nb t ng r k
4 5
4
l nb t nb r
6
5
l nb t ng r
3
6
l nb t ng r
5
7
l nb t ng r k
10 3
8
l nb t ng r
9
9
l nb t ng r k
10
l nb t ng r
11
11
l nb t ng r
12
12
l nb t ng
13
l nb t ng r
14
l nb t ng r
15
15
l nb t ng r
16
20
l nb t ng k
17
21
l nb t ng r k
14 18
22
l nb t ng k
19

Table I. Chromosome numbers in thirteen Hynobius and one Salamandrella, identified on the basis of C-banding and R-banding patterns (chromosome numbers in one longitudinal column identify whole or partially homoeologous chromosome pairs).

Species	Identified chromosome number																		Total No. of non-homoeo-logous pairs**	Total No. of iden-tified pairs	Total No. of uniden-tified pairs	Total No. of chro-mosome pairs
	Large-sized								Medium-sized						Small-sized							
H. leechii	1	2*	3	4	5	6	7	8	9	10*	11	12	13	14	15	20	21	22	0	18	10	28
tsuensis	1	2*	3	4	5	6	7	8	9	10*	11	12	13	14	15	19	20	21	0	18	10	28
dunni	1	2*	3	4	5	6	7	8	9	10*	11	12	13	14	15	19	20	21	0	18	10	28
nebulosus	1	2*	3	4	5	6	7	8	9	10*	11	12	13	14	15	20	21	22	0	18	10	28
okiensis	1	2*	3	4	5	6	7	8	9	10*	11	12	13	14	15	20	21	22	0	18	10	28
abei	1	2*	3	4	5	6	7	8	9	10*	11	12	13	14	15	20	21	22	0	18	10	28
takedai	1	2*	3	4	5	6	7	8	9	10*	11	12	13	14	15	20	21	22	0	18	10	28
tokyoensis	1	2*	3	4	5	6	7	8	9	10*	11	12	13	14	15	20	21	22	0	18	10	28
lichenatus	1	2*	3	4	5	6	7	8	9	10*	11	12	13	14	15	20	21	22	0	18	10	28
nigrescens	1	2*	3	4	5	6	7	8	9	10*	11	12	13	14	15	20	21	22	0	18	10	28
boulengeri	1	2*	3	4	5	6	7	8	9	10*	11	12	13	14	15	20	21	22	0	18	11	29
kimurae	1	2*	3	4	5	6	7	8	9	10*	11	12	13	14*	15	20	21	22	0	18	11	29
retardatus	1	–	4	6*	3*	5	7	9	10	11*	12	–	13	15	16	–	14	–	2	16	4	20
S. keyserlingii	–	–	5*	–	–	–	2	–	3*	–	–	–	–	–	–	17	18	19	14	20	11	31

*: pairs with partial homoeologous chromosomal regions
**: Identified chromosome pairs that are no longer seen in the table (H. r.: Nos. 2 and 8, S. k.: Nos. 1, 4, 6-16 and 20) possessed no homoeologous pairs in the other 13 species.

DISCUSSION

The detailed analysis conducted in the present study on chromosomes in Urodela indicated the C-bands and R-bands to be greater in number and more distinct than any other bands previously reported (Nardi *et al.*, 1973; Bailly, 1976; Rudak and Callan, 1976; Morescalchi *et al.*, 1977; Ragghianti *et al.*, 1978; Kezer an Sessions, 1979; Schmid *et al.*, 1979; Macgregor and Horner, 1980; Sessions, 1982; Sessions *et al.*, 1982; Seto *et al.*, 1982). Of the banding patterns observed for Hynobiidae, the R-banding pattern was the reverse of that of the C-banding but similar to the G-banding pattern in mammals, in which a number of bands have been noted on the arms of chromosomes. It is quite certain that Giemsa-positive bands in the C-banding pattern comprise constitutive heterochromatin (Arrighi and Hsu, 1971). Since heterochromatin is virtually inactive, variation in the size of C-positive and R-negative bands on chromosomes 2 in the 12 *Hynobius* species, and those of homoeologous chromosomes 10 in the same 12 species and chromosome 11 in *H. retardatus*, may possibly not be essential for specific differentiation.

Analysis by the conventional Giemsa method indicated most salamanders belonging to the genus *Hynobius* to exhibit essentially the same chromosome numbers, 2n=56 and 58, and to have quite similar karyotypes, though minor dissimilarities do exist (Morescalchi *et al.*, 1979; Ikebe and Kohno, 1979a, 1979b; Seto *et al.*, 1983, 1988; Seto and Matsui, 1984; Matsui *et al.*, 1985; Ikebe *et al.*, 1986). *H. retardatus* has been reported to possess a smaller chromosome number, 2n=40, and its chromosome constitution to considerably differ from those of other *Hynobius* (Azumi and Sasaki, 1971). It thus appears reasonable to conclude that nearly all salamanders belonging to the genus *Hynobius* are phylogenetically very close, with *H. retardatus* being somewhat remote from the other species in this genus. However, from studies conducted using conventional Giemsa staining, it is quite difficult to determine the relationships between species; their karyotypes differ quite considerably, as in the case of the 12 *Hynobius* species, *H. retardatus* and *S. keyserlingii*.

The banding patterns in these 14 species were compared and the percentage of regions identical among all the species was calculated. Among the 12 species, *Hynobius leechii*, *H. tsuensis*, *H. dunni*, *H. nebulosus*, *H. okiensis*, *H. abei*, *H. takedai*, *H. tokyoensis*, *H. lichenatus*, *H. nigrescens*, *H. boulengeri* and *H. kimurae*, this percentage was about 90%. In the 13 species including *H. retardatus*, it had been reduced to about 65%. This difference would not be due to that in the amount of the chromatin, since total length of chromosomes among the 13 species was in nearly all cases equal at the same condensation. It thus follows that, should the phylogenetical distance be reflected directly from identification of the banding patterns, the 12 species ought to be quite close to each other with *H. retardatus* being only somewhat remote phylogenetically. The karyotype of *H. retardatus* would thus appear to be derived from a karyotype such as found in one of the other 12 species, through chromosome fission, fusion, and/or inversion. Observation of synapses of meiotic chromosomes in interspecific hybrids should provide some clarification of chromosome similarity among *Hynobius*.

In all 14 species including *S. keyserlingii*, homoeology further decreased to about 15% (only 4 homoeologous chromosome pairs including 2 partially homoeologous). Still, the presence of the homoeologous chromosomes between these two genera, *Hynobius* and *Salamandrella*, is definite karyological evidence that species differentiation has occurred from a common ancestor.

ACKNOWLEDGEMENTS: The authors wish to gratefully acknowledge the following people for their assistance in collecting the material for use in the present study: H. Murayama, Y. Akita, the late T. Yamamoto, M. Itoh, H. Y. Lee, S. Y. Yang, N. Akaba, M. Miyamoto, K. Numata, M. Chikami, S. Tamai, M. Hasegawa, M. Matsui, K. Miyazaki, T. Takeda, M. Kawada and Y. Misawa. Appreciation is also expressed to those who provided helpful assistance at various stages of this study: T. Ohhashi, H. Yamada, K. Aoki, H. Kodama, S. Kusada, R. Katakura, Y. Izumisawa and S. Namba.

REFERENCES

Arrigi, F.E. and Hsu, T. C. (1971) Cytogenetics 10,81-86.

Azumi, J. and Sasaki, M. (1971) Chromosome Information Service 12,31-32.

Bailly, S. (1976) Chromosoma 54,61-68.

Brame, A.H., Jr. II (1985) In: Amphibian species of the world: a taxonomic and geographic reference (D.R. Frost Ed) Allen Press and the Association of Systematics Collections, pp. 562-568.

Duellman, W.E. and Trueb, L. (1986) Biology of Amphibians, McGraw-Hill, N.Y., pp. 461-466.

Dunn, E.R. (1923) Proc. Amer. Acad. Arts and Science 58,445-523.

Dutrillaux, B., Laurent, C., Couturier, J. and Lejeune, J. (1973) Sc. Paris 276D,3179-3181.

Dybowski, B. (1870) Verh. Zool. Bot. Ges. Wein 20,237-242.

Ikebe, C. and Kohno, S. (1979a) Chromosome Information Service 27,13-15.

Ikebe, C. and Kohno, S. (1979b) Proc. Japan Acad. 55B,436-440.

Ikebe, C., Kuro-o, M., Yamada, H. and Kohno, S. (1990) J. evol. Biol. 2, in press.

Ikebe, C., Aoki, K. and Kohno, S. (1987) Zool. Sci. 4,621-626.

Ikebe, C., Yamamoto, T. and Kohno, S. (1986) Zool. Sci. 3,109-113

Izumisawa, Y., Ikebe, C., Kuro-o, M. and Kohno, S., Experientia in press.

Kezer, J. and Sessions, S.K. (1979) Chromosoma 71,65-80.

Kohno, S., Kuro-o, M., Ikebe, C., Katakura, R., Izumisawa, Y., Yamamoto, T., Lee, H.Y. and Yang, S.Y. (1987) Zool. Sci. 4,81-86.

Kohno, S., Ohhashi, T., and Ikebe, C.(1983) Proc. Japan Acad. 59B,271-275.

Kuro-o, M. (1986) Thesis of Master of Science, Toho University, Chiba, Japan, pp. 1-40.

Kuro-o, M. (1986) Appendix in Thesis of Master of Science, Toho University, Chiba, Japan, pp. 53-62.

Kuro-o, M., Ikebe, C., and Kohno, S. (1986) Cytogenet. Cell Genet. 43,14-18.

Kuro-o, M., Ikebe, C., and Kohno, S. (1987a) Cytogenet. Cell Genet. 44,69-75.

Kuro-o, M., Ikebe, C. and Kohno, S. (1987b) Proc. Jpn. Acad. 63B, 109-113.

Levan, A., Fredga, K., and Sandberg, A.A. (1964) Hereditas 52,201-220.

Macgregor, H.C. and Horner, H. (1980) Chromosoma 76,111-122.

Matsui, M. (1987) Jpn. Jour. Herpetol. 12,50-64.

Matsui, M. and Miyazaki, K. (1984) Zool. Sci. 1,665-671.

Matsui, M., Seto, T., and Miyazaki, K. (1985) Jpn. J. Genet. 60, 119-123.

Morescalchi, A. (1973) In: Cytotaxonomy and Vertebrate Evolution (A.B. Chiarelli and E. Capanna, Eds) Academic Press, London and N.Y., pp. 240-258.

Morescalchi, A. (1975) Evol. Biol. 8,339-387.

Morescalchi, A., Odierna, G., and Olmo, E. (1977) Experientia 33,

1579-1581.
Morescalchi, A., Odierna, G., and Olmo, E. (1979) Experientia 35, 1434-1435.
Nakamura, K. and Ueno, S.-I. (1963) Japanese Reptiles and Amphibians in Colour [in Japanese], Hoikusha, Osaka, pp. 3-15.
Nardi, I., Ragghianti, M., and Mancino, G. (1973) Chromosoma 40, 321-331.
Perry, P. and Wolff, S. (1974) Nature 251,156-158.
Ragghianti, M., Bucci-Innocenti, S., and Mancino, G. (1978) Caryologia 31,243-256.
Rudak, E. and Callan, H.G. (1976) Chromosoma 56,349-362.
Sato, I. (1943) A Monograph of the tailed Batrachians of Japan [in Japanese], Nippon Publ. Co., Osaka, pp. 21-413.
Schmid, M., Olert, J., and Klett, C. (1979) Chromosoma 71,29-55.
Sessions, S. K. (1982) Chromosoma 84,599-621.
Sessions, S.K.; Leon, P. E., and Kezer, J. (1982) Chromosoma 86, 341-357.
Seto, T. and Matsui, M. (1984) Experientia 40,874.
Seto, T., Matsui, M. and Kawakami, E. (1986) Jpn. Jour. Herpetol. 11,137-144.
Seto, T., Matsui, M. and Kawakami, E. (1988) Jpn. Jour. Herpetol. 12,142-146.
Seto, T., Utsunomiya, Y., and Utsunomiya, T. (1982) Jpn. J. Genet. 57,527-534.
Seto, T., Utsunomiya, Y., and Utsunomiya, T. (1983) Proc. Japan Acad. 59B,231-235.
Sumner,A.T. (1972) Exptl. Cell Res. 75,304-306.
Thorn, R. (1968) Les Salamandres d'Europe d'Asie et d'Afrique du Nord. Paul Lechevalier, Paris, pp. 72-76.
Wang, X., Fang, J. and Tang, X. (1983) Acta Herpetol. Sinica 2, 19-22.

Cytogenetics of Amphibians and Reptiles
Advances in Life Sciences

NEWTS AND TWO STUDIES IN MOLECULAR CYTOGENETICS

HERBERT C. MACGREGOR

Department of Zoology, University of Leicester, Leicester, LE1 7RH, England.

SUMMARY: The current status of the taxonomy of the genus *Triturus* (Urodela; Salamandridae) is reviewed. One extraordinary chromosome and one highly repeated DNA sequence are then considered in detail in order to illustrate how clues to the evolution of European newts can be uncovered through cytogenetic and molecular studies.

Chromosome 1 in newts belonging to the cristatus species group is regularly heteromorphic. Homomorphism for this chromosome leads to embryo arrest and loss of 50% of all offspring from each single pair mating. The chromosome 1 embryo arrest syndrome has persisted for at least 10 million years. A hypothesis for its evolution is presented.

A centromeric satellite DNA that represents 10% of the genome of *T. karelini* is described. The sequence is apparently highly conserved within the genus. The age of the sequence, its distribution in the genus and its possible roles in chromosome evolution and behaviour are discussed.

INTRODUCTION

Newts of the genus *Triturus* (family *Salamandridae*) are all pond dwellers with a biphasic life cycle, including aquatic larvae and metamorphosis. The genus forms a monophyletic group characterised by a complex courtship behaviour, total absence of amplexus and a higher level of sexual dimorphism than any other amphibian genus (Halliday 1977). *Triturus* species are widely distributed in Europe and neighbouring parts of Asia (Macgregor et al. 1989). The genus is old, fossils having been recovered from the Pliocene in Spain and Poland and from the upper Miocene in France (Estes & Hoffstetter 1976, Hodrova 1985, Sanchiz 1977, Sanchiz & Mlynarski 1979, Sanchiz & Szyndlar 1984). Fossil evidence suggests that the major species groups in the genus may have been established as long as 16 million years ago.

Table 1. Taxa recognised within the genus *Triturus*, as proposed by Macgregor et al. (1989).

- subgenus *Triturus*
  - *cristatus* species group
    - *cristatus* superspecies
      - *T. carnifex*
      - *T. cristatus*
      - *T. dobrogicus*
      - *T. karelini*
    - *T. marmoratus marmoratus*
    - " " *pygmaeus*

  LARGE NEWTS

  - *T. vittatus vittatus*
  - " " *ophryticus*
  - *T. alpestris alpestris*
  - " " *apuanus*
  - " " *cyreni*
  - " " *montenegrinus*
  - " " *serdarus*
  - " " *veluchiensis*
- subgenus *Palaeotriton*
  - *T. boscai*
  - *T. italicus*
  - *vulgaris* species group
    - *T. helveticus*
    - *T. montandoni*
    - *T. vulgaris vulgaris*
    - " " *ampelensis*
    - " " *graecus*
    - " " *kosswigi*
    - " " *lantzi*
    - " " *meridionalis*
    - " " *schmidtlerorum*

  SMALL & MEDIUM SIZED NEWTS

Currently 12 species are recognised (Frost 1985, Macgregor et al. 1989). The most recent taxonomic groupings within the genus have been based on karyological and hybridisation studies of Bucci-Innocenti et al. (1983) and an analysis of mitochrondrial DNA by Wallis & Arntzen (1989). The sub-grouping of species within the genus is still a matter of debate, but certain matters are clear. The main principle in the context of this paper is that certain species can clearly be assigned to separate subgenera (Table 1). The small bodied species *T.vulgaris*, *T.helveticus* and *T.montandoni* constitute the subgenus *Palaeotriton*. The large newts of the cristatus species group, *T. cristatus*, *T.carnifex*, *T.dobrogicus*, *T. karelini* and *T. marmoratus* belong to the subgenus called *Neotriton* by Bolkay (1928), redesignated subgenus *Triturus* by Macgregor et al., (1989). Problematic species whose relationships have yet to be resolved include *T. alpestris*, *T.vittatus*, *T.boscai*, and *T.italicus*. A consensus tree based on most of the presently available phenetic and phylogenetic data has been proposed by Macgregor et al (1989) and is shown in figure 1.

Fig. 1

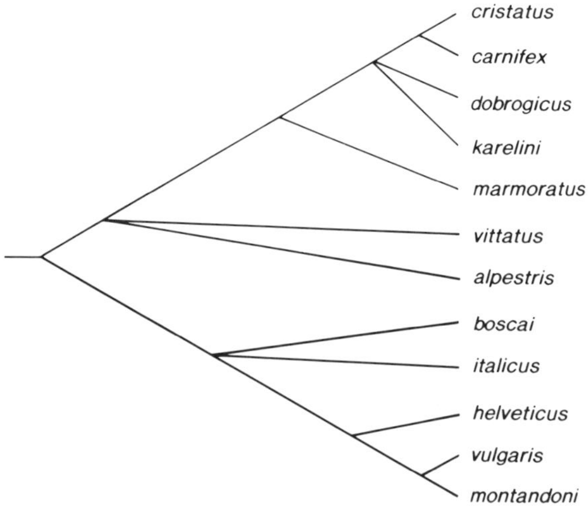

Consensus evolutionary tree of the genus Triturus based on phenetic and phylogenetic data, after Macgregor et al. 1989.

Triturus has been a focus of attention for cytogeneticists for a very long time indeed. Notable among the earliest studies were the experiments of White (1946), Lantz (1947), Lantz & Callan (1954), Callan & Spurway (1951) and Spurway & Callan (1960), all of whom were concerned with the meiotic consequences of hybridisation between different species. These investigations were valuable from several standpoints. They demonstrated possible levels of interspecific hybridisation, they provided good characterisations of the karyotypes of all the species involved and they helped towards an understanding of the chromosomal changes that have accompanied speciation within the genus. Perhaps most significantly from the standpoint of future developments in the field, they demonstrated what a truly excellent animal Triturus is for cytogenetic studies and they led on to Callan's outstanding investigation of the lampbrush chromosomes of T. cristatus and its related species (Callan & Lloyd 1960).

Essentially, the usefulness of Triturus in cytogenetics lies in its large genome and correspondingly large chromosomes, the ease with which its mitotic and meiotic chromosomes can be studied and the fact that, for reasons which can only be regarded as fortuitous, its oocytes have quite the most spectacular and easily workable lampbrush chromosomes of all organisms; although those who have worked with the lampbrush chromosomes of Pleurodeles or Notophthalmus might wish to debate the last of these claims. The study of lampbrush chromosomes in its turn opened up quite exceptional opportunities for studying the organisation and expression of specific gene sequences on chromosomes, and this provided a strong incentive to probe the molecular biology of the genus. Accordingly, we now have at our disposal a range of information, and the possibility of obtaining much more information, such as is not available for many other vertebrates.

In what follows I shall concentrate on just one chromosome and one DNA sequence in order to illustrate the usefulness of the kind of integrative biology that has proved necessary and possible in searching for clues to the evolution of European newts.

CHROMOSOME 1

All members of the cristatus species group exhibit a peculiar chromosome heteromorphism involving the long arm of the longest chromosome in the karyotype, chromosome 1 (Macgregor & Horner 1980). The long arms of the heteromorphic representatives of chromosome 1, 1A and 1B, have been shown to differ from one another with respect to length, patterns of Giemsa C-staining and the distributions of lampbrush loops with distinctive morphologies (Callan & Lloyd 1960, Sims et al. 1984). The heteromorphism is more pronounced in *T. marmoratus* than in other members of the cristatus group (Sims et al. 1984). None of the other species of *Triturus* show heteromorphism for chromosome 1.

Despite suggestions prior to 1975 and some more recent arguments (Wallace 1984, 1987), the heteromorphism of chromosome 1 is not related to sex determination. It is present in all individuals, males and females (Morgan 1978) and a distinct XY sex chromosome pair has been described in *T. alpestris* (chromosome 4) and *T. vulgaris* (chromosome 5) by Schmid et al. (1979), and in all members of the cristatus species group (chromosome 4) by Sims et al. (1984).

Chromosome 1 in the cristatus species group appears to represent a balanced lethal system in which only 1A/1B combinations are viable. Embryos that are homomorphic 1A/1A or 1B/1B, representing 50% of the expected zygotes produced by each mating, stop developing and die at the late tailbud stage of embryogenesis. The development of embryos homomorphic for the shorter of the 2 chromosomes, 1B/1B, seems to be primarily a consequence of failure to complete gastrulation, and many of the features of these embryos can be explained by an abnormal arrangement of endoderm caused by incomplete epiboly. The development of 1A/1A embryos is more normal up to the arrest at late tailbud (Sessions et al. 1988)

The chromosome 1 embryo arrest phenomenon appears to be a uniquely derived characteristic that supports the concept that all the large bodied newts, T.cristatus, T. carnifex, T.karelini, T.dobrogicus and T. marmoratus are members of a monophyletic lineage, as originally proposed on osteological grounds by Bolkay (1928).

It seems certain that the developmental arrest phenomenon has played an important role in the evolutionary biology of Triturus. The arrest syndrome has persisted for 10-12 million years (Sessions et al. 1988) despite the substantial reproductive failure that it generates. Indeed the high level of reproductive failure stands in sharp contrast to the reproductive success of several other species of Triturus that are commonly sympatric with one or more of the cristatus species. How, one must ask, has the cristatus group of species survived for so long with a genetic defect that regularly costs them 50% of their offspring, sharing ecosystems with other newts that do not suffer this handicap. The question is undoubtedly an ecological one that should be addressed with attention to matters such as competition between different species occupying the same pond, feeding habits of larval and adult newts, numbers of eggs deposited, times of spawnings and rates of development. On the other hand it could be said that the fitness of each individual within a given population is equally affected and intraspecific natural selection cannot therefore take place. However, it seems likely that this particular situation is more complex and perhaps more interesting.

THE CYTOLOGY OF CHROMOSOME 1

The relative length of T. carnifex chromosome 1A, at 11.02, is greater than that of any other chromosome in the genus except for its counterpart in T. marmoratus (Sims et al. 1984). The chromosome must therefore have grown since the cristatus species group diverged from a common Triturus ancestor, and most of the growth must have taken place in the long arm. The chromosomes 1 of T. marmoratus are both much more metacentric than those of carnifex, (Figure 2) yet the chromosomes 1A of both species are about the same overall size and the chromosome 1B of marmoratus is actually slightly smaller than its counterpart in carnifex. The shift in centromere position that accompanied speciation is therefore more likely to have been due to a structural rearrangement than to growth of the long arms of the marmoratus chromosomes. Asynapsis is confined to the heterochromatic (Giemsa C-positive) region in T.carnifex, but extends well beyond this region and right across the centromere into the short arm in marmoratus.

One of the most curious features of the heteromorphic region of chromosome 1 is its variability. Indeed, not only is it curious, but it has proved to be a real source of confusion in

Fig. 2

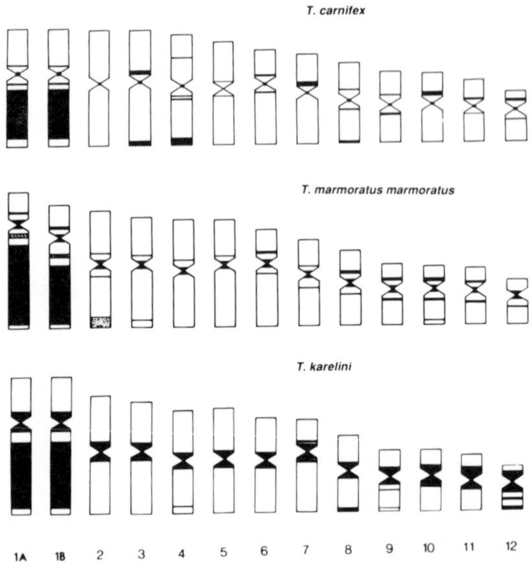

Giemsa C-band karyotypes for T. carnifex, T. karelini, and T. marmoratus. Note that the chromosomes 1 of T. marmoratus are more sub-metacentric than those of carnifex or karelini, the chromosome 1B of marmoratus is shorter than 1A, and T. karelini is represented as having large centromeric C-bands as compared with the other two species.

evaluating certain other properties of the chromosome and in making comparisons between observations made by different investigators in different parts of the world. Every species (earlier known as sub-species or races) in the cristatus species group has a different chromosome 1 C-band pattern and a different arrangement of distinctive lampbrush loops in this region. Moreover, there is widespread variation in the chromosome 1 C-band pattern between populations of the same species. The carnifex C-bands shown by Macgregor & Horner (1980), for example, are quite different from those shown some years later in the same species by Sims et al. (1984). The C-band patterns published for carnifex are quite different from those of cristatus (Sessions et al. 1988). In the latter paper we specifically drew attention to this matter and pointed out the importance of confining future investigations on chromosome 1 heteromorphism to animals from defined populations (=ponds) so as to avoid complications that might arise through misinterpretation of homo- versus hetero-morphism.

Variability in chromosome 1 goes even deeper. There is widespread variation WITHIN populations in the locations on the heteromorphic arms of lampbrush loops that in-situ hybridise with specific repetitive DNA probes (Varley et al. 1980a,b, Morgan et al. 1980, Macgregor et al. 1981).

The molecular biology of chromosome 1 must hold all kinds of interesting clues with regard to the arrangement, behaviour and expression of repetitive DNA sequences. It was on this chromosome that the first definitive evidence of the transcription of highly repetitive 'satellite' DNA was found (Varley et al. 1980a,b). Two specific satellites were identified. One of them TcS1 is 390 base pairs in length and represents about 1% of the genome of T. carnifex. It is present in relatively large amounts in the heteromorphic regions of chromosome 1, and it is also widely dispersed in all the other chromosomes of the set. It is transcribed only in the oocyte nucleus and the transcripts are confined to the nucleus. It is probably one of several highly repeated sequences that have become amplified in the long arm of chromosome 1 and are transcribed only on lampbrush loops. Transcription is thought to be a consequence of non-stop read-through by RNA polymerase II that initiates at functional gene promoters but continues to transcribe repetitive DNA that lies downstream from the ends of functional genes (Varley et al. 1980a, b, Epstein et al. 1986).

Much of the DNA in the long arm of chromosome 1 is A+T rich and seems to contain representatives of virtually every DNA sequence, coding or non-coding, that has been in-situ hybridised to it, with the notable exception of a highly abundant and evolutionarily conserved centromeric satellite isolated from T. karelini (=T.c. karelinii) (Baldwin & Macgregor 1985). The long arm of the chromosome 1 is undoubtedly a very strange piece of chromatin, and we can only suppose that it has become so through structural isolation, exposure to all the normal forces of genome change, and freedom from the rigours and checks of the meiotic process.

In more general terms, it is essential to understand the scale of the molecular scenario on chromosome 1 - a matter that is all too often overlooked by persons who are unaccustomed to working with the large genomes of urodeles. There are 21 pg of DNA (about 2.4×10^{10} base pairs) in the haploid genome of a crested newt (Macgregor et al. 1989). Of this about 1.3 pg (1.2×10^{9} base pairs) make up the heteromorphic region of chromosome 1, which is equivalent to about one third of the entire haploid genome of man.

THE EVOLUTION OF CHROMOSOME 1

There have been a number of attempts to explain the evolution of chromosome 1 in newts of the cristatus species group (Macgregor & Horner 1980, Sims et al. 1984, Wallace, 1984, 1987, Sessions et al. 1988). All of these publications come from just two laboratories, the author's and that of Dr. Hugh Wallace.

In our most recent interpretation of the evidence to hand (Sessions et al. 1988) we have proposed that the chromosome 1

Fig. 3

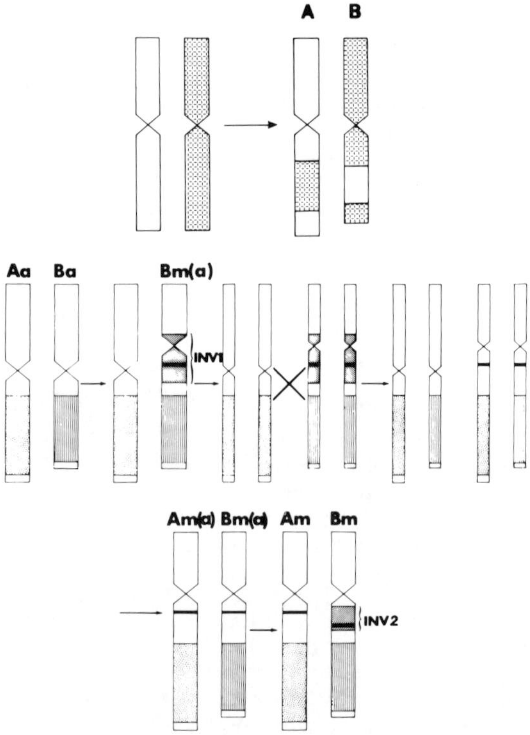

Scheme for the evolution of chromosomes 1A and 1B in the cristatus species group, showing the consequences of a large unequal exchange (top line) followed by the inversions and crossovers needed to establish the chromosomes 1 of T. marmoratus. Aa, Ba, ancestral chromosomes 1A and 1B; Am(a) and Bm(a), ancestral marmoratus chromosomes 1A and 1B; Am, Bm, present day marmoratus chromosomes; INV, inversion.

arrest syndrome is the result of a cytogenetic accident involving a large unequal exchange between the homologues of chromosome 1 in the common ancestor of the cristatus species group (T. cristatus, T. carnifex, T. karelini, T. dobrogicus, and T.marmoratus) (Figure 3). For such an exchange to result in the formation of a balanced lethal chromosomal homomorphism in one step we have to suppose that it happened in the germ line and included DNA sequences that were important for early development. Following this intial event, the long arm of chromosome 1 would be meiotically isolated, which would release constraints on the structure of the arm and permit growth by accummulation of repetitive DNA and the formation of heterochromatin.

The importance of the initial unequal exchange event is that it could generate the two alternative homomorphic (homozygous) phenotypes such as have been described here and by Sessions et al. (1988). One of the arrest phenotypes would carry both the chromosomes 1B that had the larger deletion, and it might be expected to show more severe abnormalities than embryos that were homomorphic for the chromosome 1A carrying the smaller deletion.

At some stage in the early evolution of chromosome 1 T. marmoratus diverged from the remainder of the cristatus species group (Macgregor et al. 1989) and one of the heteromorphic chromosomes 1 of marmoratus underwent a major pericentric inversion such that it became much more submetacentric. A substantial length of euchromatin now separated the centromere on the inverted chromsome from the start of the heterochromatic heteromorphic region on the long arm. Meiotic crossing over in the small segment of euchromatin distal to the limits of the inversion then produced the marmoratus chromosomes 1A and 1B both of which had their centromeres in the new submetacentric position, and in the course of time these two chromosomes were the ones that spread throughout the new marmoratus species. Much later another small paracentric inversion formed in the proximal euchromatin of the long arm of chromosome 1B in marmoratus. This had the effect of suppressing synapsis and chiasma formation throughout the whole of the long arm of chromosome 1 in T. marmoratus, except for tis extreme tip. The entire scheme is illustrated in Fig. 3. All the modern T. marmoratus that we have examined are still clearly heterozygous for this inversion (Sims et al. 1984).

The concept of an unequal genetic exchange as the basis for a balanced lethal homomorphism is supported by studies of interspecific hybridisation in Triturus. F1 hybrids have been obtained for matings between T. carnifex (cristatus species group) and T. vulgaris (Mancino et al. 1978) indicating that the effect of having only one carnifex chromosome 1 can be corrected by the presence of the equivalent 'normal' chromosome from T. vulgaris. Natural hybrids, T. cristatus x T.marmoratus, are known all to be heteromorphic for chromosome 1 and possess both possible combinations of chromosomes 1A and 1B from each species (Sims et al. 1984). Perhaps most interestingly, triploid F1 hybrids between T.cristatus and T. marmoratus survive and develop into normal adults (Lantz & Callan 1954) showing that the presence of 2 homomorphic chromosomes 1 is not in itself sufficient to produce embryo arrest, so long as one chromosome of the other type is also present.

If the first exchange event in the evolution of chromosome 1 happened as a sister chromatid exchange in an early gonial mitosis, then the altered chromosomes would be distributed among a large number gametes. A mating between an individual carrying such gametes and one with normal gametes would produce mutant/normal heterozygotes. If these survived then it would take just one more generation with sibling matings to produce the first A/B chromosome 1 heterozygotes as well as the first

A/A and B/B arresting embryos. Accordingly, all members of the cristatus species group must have evolved from a single small ancestral population that was dominated by 1A/1B heterozygotes. In essence all members of the cristatus species group must have originated in a single small pond that existed 10 or more million years ago somewhere on the land mass that has since become the continent of Europe.

THE KARELINI SATELLITE

The second part of this paper is concerned with a highly repetitive DNA sequence that was first isolated and characterised in the large southern European newt *T.karelini*. To avoid confusion it should immediately be said that in all previous reports of work on this animal it has been called *Triturus cristatus karelinii*. The change in nomenclature to *T. karelini* is based on two considerations. First, the newt is now regarded as having full species status, rather than being a subspecies of *T. cristatus* (Frost 1985, Macgregor et al. 1989). Second, it is named after the 19th Century Russian naturalist Karelin, and should therefore be named *karelini* and not *karelinii*.

Highly repetitive ('satellite') DNA sequences can be identified as discrete bands following electrophoresis on agarose gels of DNA fragments produced by treatment of whole genomic DNA with restriction enzymes. In *Triturus* this material is usually, but not always, cytologically localised in Giemsa C-positive regions of the chromosome.

In 1960, Callan & Lloyd made a detailed study of the lampbrush chromsomes of 4 members of the cristatus species group, including *T. karelini*. Naturally the location and description of the centromeres on lampbrush chromosomes was one of their major objectives. For 3 of the species they studied this was a difficult task, but they succeeded nonetheless. In *T. karelini* it was exceptionally easy. On all of the lampbrush chromosomes of this species the tiny granules that were considered to represent the definitive 'centromeres' were immediately flanked on both sides by large conspicuous 'bars' of dense chromatin that carried no lateral loops. These 'centromere bars' can be up to 5 um in length and 1-2 um wide, so that the entire centromere region on a *karelini* lampbrush chromosome is marked by a structure that is almost as large as an average mitotic metaphase chromosome and is linearly integrated into a lampbrush chromosome that may be up to half a millimeter long (Fig. 4). No such centromere bars are found in any other *Triturus* species, although similar structures have been reported in the American newt *Notophthalmus* and in some plethodontid salamanders.

Fig. 4

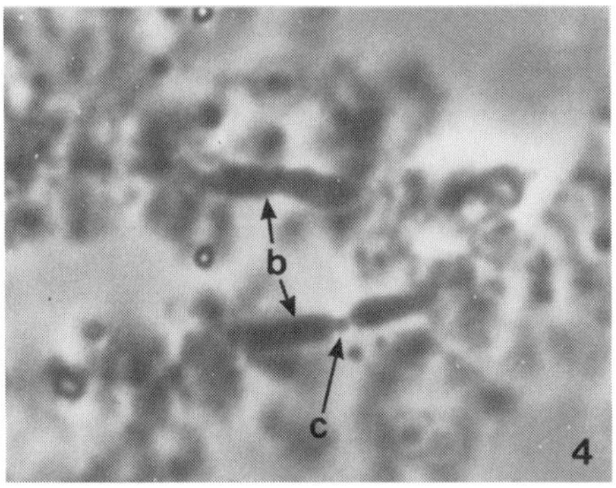

Phase contrast micrograph of the centromere region of a lampbrush from T. karelini, showing the large centromere bars (b) flanking the small centromere granule (C).

Fig. 5

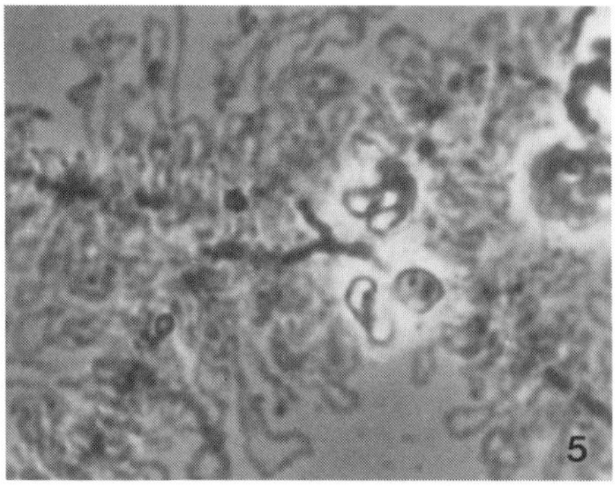

Phase contrast micrograph of part of a lampbrush chromosome from T. karelini showing fusion between two homologous sets of centromere bars to form a Y configuration.

The centromere bars are truly remarkable structures. They stain strongly with the Feulgen reaction, with ethidium bromide and with DAPI, indicating that they consist of DNP. Homologous centromere bars often fuse with one another, wholly or partially, to form a single rod or an X or Y configuration (Fig. 5). Fusion between non-homologous centromere bars has never been reported. In larger oocytes the centromere bars are usually continuous through the centromere, with the result that the centromere granule itself may be laterally displaced from the chromosome axis. The bars appear to grow during the later stages of oogenesis as a consequence of progressive incorporation of neighbouring chromomeres into their substance (Callan & Lloyd 1960).

The chromatin of centromere bars is certainly different from that of neighbouring chromomeres and from that of condensed mitotic chromosomes. They stain less intensely with DAPI than the chromomeres. They behave differently from the chromomeres when the chromosomes are exposed to trypsin (Macgregor & Callan 1962), and their surface structure, when viewed with a scanning electron microscope, is quite unlike that of other forms of condensed chromatin (Macgregor 1986).

Some years after Callan & Lloyd's description of the lampbrush chromosomes of T. karelini we looked at the Giemsa C-staining properties of Triturus mitotic chromosomes. One thing was immediately apparent. The chromosomes of T. karelini all carried centromeric C-bands that were massively large as compared to those of other newts (Fig. 2). Since C-band heterochromatin is generally an indication of the presence of highly repeated DNA, we immediately examined the DNA of _T. karelini_ and identified a low molecular weight component that appeared after digestion of whole DNA with Hae III. This 'Hae III satellite' was cloned into a bacterial plasmid vector, labelled with tritium and in-situ hybridised to lampbrush and mitotic chromosomes. The results were clear and conclusive. We had a 32/33 base pair sequence that represented up to 10% of the genome and was strictly localised in the centromeric heterochromatin of mitotic chromosomes and the centromere bars of lampbrush chromosomes. The same sequence was present and centromerically localised in the related _T. cristatus_, but in far smaller quantities (less than 1% of the genome), which was consonant with the fact that _T. cristatus_ has very small centromeric C-bands and no lampbrush centromere bars. The sequence seemed to be absent from the small bodied newts _T. vulgaris_, _T. helveticus_ and _T. alpestris_. The sequence was named TkS1. Another less abundant sequence, TkS2, also centromerically localised, was also identified (Baldwin & Macgregor 1985).

The features that we regarded as most significant in relation to TkS1 were that it was centromerically localised, it was small and very highly repeated, it was abundant in _T. karelini_ and it was present in appreciable amounts only in members of the cristatus species group.

It was suggested that amplification of this sequence in the regions of the centromeres could have come about through frequent unequal cross overs and exchanges between chromosomes that would be facilitated by the clustering and fusion of centromeric heterochromatin that is particularly common in the cells of newts (Baldwin & Macgregor 1985).

The idea that such exchanges can and do occur and lead to redistribution of heterochromatin between chromosomes is supported by some remarkable observations made by Callan in the 1960's and reported in his 1986 monograph on lampbrush chromosomes (Callan 1986). Callan found that F1 hybrids between _T. karelini_ and _T. cristatus_ were heterozygous for the presence of centromere bars on all bivalents, as might be expected. When he backcrossed the F1 females to _T. cristatus_ some of the bivalents were heterozygous for centromere bars and others did not have them at all, again as might be expected. However, although the centromere bars could easily be identified in backcross hybrids, they were often smaller and less compact than in the grandparental _karelini_ and "with certain regularity pairs of lateral loops extended from some of them. At the base of such loops constrictions appear in the axial (centromere) bars" (Callan 1982). In my view, the phenomenon observed by Callan could well have been a consequence of crossing over between centromeric regions of either parental type, so diluting or interrupting the blocks of heterochromatin that are normally present in *T. karelini*.

At the time of writing, at least 8 different satellite sequences from 3 species of *Triturus* have been isolated, purified by molecular cloning and sequenced (Varley et al. 1980a,b, Baldwin & Macgregor 1985, Barsacchi et al. 1986, Cremsi et al. 1988, Macgregor and Sessions 1986a,b). Two satellites have been isolated from *Notophthalmus viridescens* (Diaz et al. 1981, Gall et al. 1981, Epstein et al. 1986, 1987). One satellite has been isolated but not sequenced from the plethodontid salamander *Plethodon cinereus* (Macgregor & Kezer, 1971, Macgregor et al. 1973). The characteristics of the *Triturus* and *Notophthalmus* satellites are listed in Table 2.

Table 2

Satellite	Size Base Pairs	% of genome	Present in										chromosome location
			Nv	Tcr	Tk	Tc	Tm	Tvt	Ta	Tv	Th	Tb	
TcS1	390	1	X	X	X	X	X	X	X	X	X	X	dispersed
TcS2	330	0.1	X	X	X	X	X	X	X	X	X	X	dispersed
Nv2	330	0.09	X	X	?	X	?	?	X	X	?	?	dispersed
Tvm1	313	1	X	X	?	?	?	?	X	X	?	?	pericentric
TkS1	32/33	5-10	-	X	X	X	X	x	x	-	x	-	centromeric
TkS2	68	0.5-1	-	X	X	X	X	-	-	-	-	-	centromeric
Tvm7	200	1	?	-	?	-	?	?	-	X	X	?	centromeric
Nv1	222	0.6	X	?	?	-	?	?	-	-	?	?	centromeric

In the left hand column the first letter specifies the genus and the second letter the species from which the satellites were prepared. Tv, *Notophthalmus viridescens*, Tcr, *T. cristatus*, Tk, *T. karelini*, Tc, *T. carnifex*, Tm, *T. marmoratus*, Tvt, *T.vittatus*, Ta, *T. alpestris*, Tv, *T. vulgaris*, Th, *T. helveticus*, Tb, *T. boscai*. X, sequence present in amounts equivalent to at least one tenth of the level present in the species from which it was prepared; x, sequence present, but at less than one hundredth of the amount present in species from which it was prepared; -, sequence throught to be absent; ?, no information.

A good correlation exists between the supposed 'age' of a satellite sequence and its location on the chromosome set. Satellites that are present in a wide range of species and have retained the same sequence (to the extent that they readily cross-hybridise) are regarded as 'old'. Good examples of old sequences are TcS2 and Nv2 (Table 1). Satellites that are present in only one species or a closely related group of species are regarded as 'young'. We shall see later that if the term 'young' is to have any substance whatever in this argument, then it must be applied very much in the relative sense. TkS1 is a good example of a young sequence. Evidence from in-situ hybridisations has shown that of the sequences so far investigated, old satellites tend to be widely dispersed over the entire chromosome set. Young satellites, like TkS1 and Tvm7 tend to be more or less strictly localised in centromeric C-bands, and satellites of intermediate age occur in pericentric or interstitial C-bands. On this basis we proposed a model for chromosome evolution (Macgregor & Sessions 1986a,b) in which karyotypes experience cycles of amplification of tandem arrays of satellite DNAs usually commencing at the centromeres (or telomeres). These arrays form part of the material of centromeric C-bands. The arrays then break up and become dispersed away from the original sites of amplifcation by a series of structural rearrangements affecting all chromosomes, and there is a gradual divergence of the original satellite sequence through base substitutions, additions or deletions. Sequence divergence eventually leads to the disappearance of the satellite in that it is no longer detectable as a band after restriction enzyme digestion and gel electrophoresis. If the rate of dispersal exceeds that of sequence divergence, such that smaller but still homogeneous tandem arrays persist, then C-bands will become smaller but will still be detectable in pericentric or interstitial positions. As an adjunct to our model we proposed that the commonest positions for large blocks of C-band heterochromatin are at the telomeres and centromeres, and this may be taken to indicate that in these regions there is the greatest propensity for the kinds of physical and molecular associations between chromosomes that may be expected to lead to the highest levels of inter-chromatid exchange such as would quickly generate and homogenise tandem repeats (Smith, 1973, 1976).

Our model can be applied to newts of the genus _Triturus_ and we think it may perhaps be applicable to a wide range of organisms. The model may also be used as a basis for an explanation for evolutionary changes in genome size that involve balanced growth of chromosomes without change in chromosome number or shape. Good examples of this kind of phenomenon are to be found in _Plethodon_ (Mizuno & Macgregor 1974), in the crustacean _Gammarus_, in the turbellarian _Mesostoma_ and in the angiosperm plant _Crepis_ (Callan 1967).

Cremsi et al. (1989) recently made an another suggestion. They proposed that a sequence such as that in Tvm7 (centromeric and confined to the Palaeotriton group of newts) may have been present in the ancestor to all present day Triturus species. The sequence has diverged over about 20 million years of speciation within the genus. It is found in T. vulgaris and its closest relative within the sub-genus Palaeotriton, T. helveticus, simply because these have become distinct species relatively recently. Absence of molecular cross-hybridisation of the sequence to DNA in other more phylogenetically distant species could be explained by more or less extensive sequence divergence. One inference that can be drawn from this suggestion is that the cytological location of a satellite sequence is of little significance.

Cremsi et al. (1989) went on to point out that according to the model of Macgregor & Sessions (1986a, b), one would expect a closer inter-species sequence homology between repeats of a centromeric sequence than between repeats of a pericentric or dispersed one. Accordingly they suggest that repeats of the same centromeric family taken from different species should show extensive sequence variation even though they should still possess sufficient homology to enable cross-hybridisation. Repeats of pericentric or dispersed families should display a relatively high degree of homology.

Table 3

Type 1

CCAGAGTAAGAGT TAAGACAGTAGCTATGAGG 32 bases

Type 2

CCAGAGTAAGAGT TAAGACCTTAACCATTAGG 32 bases

Type 3

CCAGAGTAAGAGTCTAAGACCTTAACTATTAGG 33 bases

Consensus sequences for the three types of TkS1. The gap at position 14 (left to right) in types 1 and 2 corresponds to the position of the extra base that characterises type 3. (After Varley and Macgregor 1989).

We have recently carried out some studies of the highly repeated centromeric sequence TkS1 through which we hoped to resolve this matter (Varley & Macgregor 1989). We have prepared and sequenced 101 monomer clones of TkS1 from T. karelini (3 individuals from 3 separate populations), 43 monomer clones from one individual of T. cristatus, 25 monomer clones from one individual of T. carnifex, 22 monomer clones from T. dobrogicus and 17 monomer clones from T. marmoratus. The sequences of these clones are all closely similar. The clones fall into 3 clearly distinguisable types. Type 1 consists of 32 bases, and of the total of 105 representatives of this type 32 show single base substitutions all of which are at positions that are apparently random. Type 2 consists of 32 bases, and it differs from type 1 with respect to just 5 bases located in a region of 10 bases towards the right hand end of the sequence. The remainder of the sequence is identical to the consensus sequence of type 1. Type 3 consists of 33 bases and is closely similar to type 2 apart from the presence of an extra base in position 14, reading from the left to right. The consensus sequences for the three types are shown in Table 3. A minority of the sequences of types 2 and 3 also have single base substitutions at apparently random positions.

Table 4

Species	% Frequences			No. of clones
	Type 1	Type 2	Type 3	
T. karelini	55	22	23	101
T. cristatus	27	43	30	25
T. carnifex	12	56	32	44
T. dobrogicus	9	73	18	22
T. marmoratus	0	47	53	17

All three types of TkS1 are present in all of the 5 species so far examined, but the frequencies with which they are represented in our samples vary from one species to another (Table 4).

In relation to the alternative hypothesis outlined by Cremsi et al. (1988) it is therefore possible to say at this stage that TkS1, a sequence that is exclusively centromerically localised, has undergone changes that have led to 2 divergences, one involving 5 bases and the other an addition of a single base. The divergent sequences resulting from these changes are present in all 5 species from which we may conclude that they all happened before divergence of the species belonging to the cristatus group. The matter of differing relative frequencies of sequence types is something that requires further investigation.

In straightforward terms it would seem that the level of homology within TkS1 across all 5 members of the cristatus species group is high, which, following the argument of Cremsi et al. (1988) would seem not to conflict with the notion that TkS1 is relatively 'young', so leaving some substance in the sequence dispersal model for chromosome evolution proposed by Macgregor and Sessions (1986a and b). However, that is not, of course, the end of the story. Our studies of the TkS1 sequence have presented us with a few surprises and they have opened up some new questions. How, by any stretch of the imagination, can a sequence that has remained substantially unchanged for more than 10 million years be described as 'young'? Why is it that 2 of the TkS1 types are found in all 5 species and all 3 types are found in 4 of these species? Why is it that out of the 208 clones that we have sequenced only one carries a single base substitution in the first 13 nucleotides? In all others these 13 nucleotides are identical, and have remained so for a very long time indeed. Why is the TkS1 so strictly localised to the centromeric regions of all chromosomes? All of the other repetitive DNAs that have been examined in Triturus are either widely dispersed AND centromeric, and/or concentrated in the heterochromatin of the long arm of chromosome 1 - a region that seems to have acted as a kind of 'black hole' for all manner of nucleotide arrays - except TkS1.

Let it be accepted for the moment that one of the major forces in generating large tandem arrays of sequences is unequal sister chromatid exchange (Smith 1976, Stephan 1987, 1989) and that the opportunities for this are likely to be high in heterochromatin that is commonly fused. With regard to TkS1 there would seem to be 3 possibilities: (1) all 3 types of the sequence are liberally intermixed as monomers or small multimers in all parts of the whole array on each chromosome; (2) all 3 types are intermixed as large tandem arrays on all chromosomes; (3) each type is found exclusively on particular chromosomes.

The third of these arrangements seems unlikely, but it is testable by in-situ hybridisation, using the right hand end of the sequence as a probe and stringent washing procedures. The experiment will eventually be done and the matter resolved.

The second arrangement seems equally unlikely. Every event of unequal sister chromatid exchange will remove arrays of repeated sequences from a chromosome as well as duplicate them. This being so, if at one point in time an organism has 3 types of a repeated sequence and these are arranged in large linked tandem arrays on the same chromosome, then the likelihood is that one or more of these types will eventually be eliminated. This has not happened to TkS1.

The most likely arrangement is a more or less random mixing of monomers or short arrays of all three types in all chromosomes. This, again, can and will be tested by further sequencing studies and by taking advantage of the fact that types 2 and 3 can be distinguished from type 1 by an Alu1 restriction site in the right half of the sequence.

With regard to the substitutions that have been identified in TkS1, there are few surprises. Assuming a selectively neutral sequence, a random mutation rate of between 1 and 8 x 10^{-9} per base per year (Li et al. 1985) and an estimated divergence time for T. marmoratus and other members of the cristatus species group of 16 million years (real time evolution of 32 million years), we would expect to find between 3% and 25% base substitutions in the 32 base pair monomer. The observed incidence of substitutions is within this range for the whole sequence and also for the right hand 20 bases of the sequence. On the other hand, the first 13 bases of the sequence have remained unchanged.

For the past two decades it has been customary to regard short sequence highly repetitive DNA as having no function and being selectively neutral. As a scarred veteran who withdrew from this bloody field a few years ago (Macgregor & Sessions 1986a), I am naturally hesitant about any re-engagement. Nevertheless new evidence is coming forward and there is a clear revival of interest in satellites. The catalysts for a rethink have, in my view, been the discovery of hypervariable DNA sequences (Jeffreys 1985),the discovery of short non-coding repeat sequences that are esssential for the roles of telomeres and centromeres (Bloom et al. 1986, Blackburn 1984), and a general recognition of the fact that DNA sequences can have roles besides those associated with transcription and translation of proteins, roles than involve specific interactions with proteins (Fanning, 1987, Wong & Rattner 1988, Masumoto et al. 1989), patterns of condensation of chromatin (Radic et al. 1987) and associations between different cellular components (Neuer-Nitsche et al. 1988). It is outwith the topic of this symposium to discuss such fundamental molecular matters, but since highly repeated DNA is particularly abundant in the genomes of urodeles, two aspects of the situation are perhaps worth a mention.

First, selective neutrality may now be a slightly dangerous and facile assumption to make with regard to satellite DNA, and whether or not we make this assumption will undoubtedly condition our view of the usefulness of highly repeated DNA in phylogenetic studies. Recent articles by Stephan (1987, 1989) merit attention in this regard. One of his conclusions (Stephan 1987), for example, is that small populations will have, on average, more repetitive DNA in their genomes than large ones. How, we might ask does this fit with everything we know about repetitive DNA in newts and salamanders. The DNA has been examined and the population biology is surely known, but the two aspects have never been considered alongside one another.

Secondly, and with specific regard to TkS1, we should remain alert for unusual features of satellites that may provide clues to special functions. Why for example, have 13 bases of TkS1 remained unchanged? Why has the sequence not 'leaked' into the long arm of chromosome 1? Is it perhaps selectively disadvantageous to have this particular sequence in locations other than centromeric, and if so then are we not half way to knowing that TkS1 has something to do with centromeres?

REFERENCES

Baldwin, L., Macgregor, H.C. (1985). Centromeric satellite DNA in the newt _Triturus cristatus karelinii_ and related species: its distribution and transcription on lampbrush chromosomes. Chromosoma (Berl.) _92_ 100-107.

Barsacchi-Pilone, G., Batistoni, R., Andronico, F., Vitelli, L., Nardi, I. (1986). Heterochromatic DNA in _Triturus_ (Amphibia, Urodela). I.A satellite DNA compound of pericentric C-bands. Chromosoma (Berl.) _93_, 435-446.

Blackburn, E.H. (1984). Telomeres: do the ends justify the means? Cell _37_ 7-8.

Bloom, K., Hill, A., Yeh, E. (1986). Structural analysis of a yeast centromere. BioEssays _4_, 100-104.

Bolkay, St. J. (1928). Die Schnadel der Salamandrinen, mit besonderer Rucksicht auf ihre systematische Bedeutung. Zeits.Anat.Entwicklungsgesch. _86_, 259-319.

Bucci-Innocenti, S., Ragghianti, M., Mancino, G. (1983). Investigations of karyology and hybrids in _Triturus boscai_ and _T. vittatus_, with a reinterpretation of the species groups within _Triturus_ (Caudata:Salamandridae). Copeia 1983, 662-672.

Callan, H.G. (1967). The organisation of genetic units in chromosomes. J. Cell Sci. _2_, 1-7.

Callan, H.G. (1986). Lampbrush Chromosomes. Springer Verlag, Berlin, Heidelberg, New York, Tokyo.

Callan, H.G. (1982). Lampbrush chromosomes. Proc.R.Soc. Lond. 214, 417-448.

Callan, H.G. and Lloyd, L. (1960). Lampbrush chromosomes of crested newts _Triturus cristatus_ (Laurenti). Phil.Trans.R.Soc.Lond. B243, 135-212.

Callan, H.G., Spurway, H. (1951). A study of meiosis in interracial hybrids of the newt _Triturus cristatus_. J.Genet._50_. 235-249.

Cremsi, F., Vignali, R., Batistoni, R., Barsacchi-Pilone, G. (1988). Heterochromatic DNA in _Triturus_ (Amphibia, Urodela). II.A centromeric satellite DNA. Chromosoma (Berl.) _97_, 204-211.

Diaz, M.O., Barsacchi-Pilone, G., Mahon, K.A., Gall, J.G. (1981). Transcripts from both strands of a satellite DNA occur on lampbrush chromosome loops of the newt _Notophthalmus_. Cell _24_, 649-659.

Epstein, L.M., Mahon, K.A., Gall, J.G. (1986). Transcription of a satellite DNA in the newt. J.Cell Biol. _103_ 1137-1144.

Estes, R. and Hoffstetter, R. (1976). Les urodeles du Miocene de la Grive-Saint-Alban (Isere, France). Bull.Mus.Natn.Hist.Nat.Parish (Sci.Terre 398) _57_ 299-343.

Fanning, T.G. (1987). Origin and evolution of a major feline satellite DNA. J.Mol.Biol._197_, 627-634.

Frost, D.R. (ed.) (1985). Amphibian species of the World: a taxonomic and geographical reference. Allen Press, Association of Systematics Collections, Lawrence, Kansas.

Gall, J.G., Stephenson, E.C., Erba, H.P., Diaz, M.O., Barsacchi-Pilone, G. (1981). Histone genes are located at the sphere loci of newt lampbrush chromosomes. Chromosoma (Berl.) 84, 159-171.

Halliday, T.R. (1977). The courtship of European newts: an evolutionary perspective. In The Reproductive Biology of Amphibians (eds. D.H. Taylor & S.I. Guttman). Plenum Press, New York. pp. 185-232.

Hodrova, M. (1985). Amphibia of Pliocene and Pleistocene Vcelare localities (Slovakia). Casopis Minerologii Geologii 30, 145-161.

Jeffreys, A.J., Wilson, V., Thein, S.L. (1985). Hypervariable 'minisatellite' regions in human DNA. Nature (Lond.) 314, 67-73

Lantz, L.A. (1947). Hybrids between Triturus cristatus Laur. and Triturus marmoratus Latr. Proc. Zool. Soc. (Lond.) 117, 247-258.

Lantz, L.A., Callan, H.G. (1954). Phenotypes and spermatogenesis of interspecific hybrids between Triturus cristatus and T. marmoratus. J. Genet. 52, 165-185.

Li, W.-H., Luo, C. -C., Wu, C. -I. (1985). In Molecular Evolutionary Genetics (MacIntyre, R.J. ed.). pp. 1-94. Plenum Press, New York.

Macgregor, H.C. (1986). The lampbrush chromosomes of animal oocytes. In Chromosome Structure and Function (ed. M.S. Risley). Van Nostrand Reinhold Company, New York. pp. 152-186.

Macgregor, H.C., Calan, H.G. (1962). The actions of enzymes on lampbrush chromosomes. Quart. J.Microsc. Sci. 103, 173-203.

Macgregor, H.C., Horner, H.A. (1980). Heteromorphism for chromosome 1, a requirement for normal development in crested newts. Chromosoma (Berl.) 76, 111-122.

Macgregor, H.C., and Kezer, J. (1971). The chromosomal localisation of a heavy satellite DNA in the testes of Plethodon c. cinereus. Chromosoma (Berl.) 33, 167-182.

Macgregor, H.C., Sessions, S.K. (1986a). The biological significance of variation in satellite DNA and heterochromatin in newts of the genus Triturus: an evolutionary perspective. Phil.Trans.R.Soc.Lond. B 312, 243-259.

Macgregor, H.C., Sessions, S.K. (1986b). Models for evolution in large genomes and karyotypes of urodeles. Verh.Dtsch.Zool.Ges. 79, 137-148.

Macgregor, H.C., Horner, H., Owen, C.A. and Parker I. (1973). Observations on centromeric heterochromatin and satellite DNA in salamanders of the genus Plethodon. Chromosoma (Berl.) 43, 329-348.

Macgregor, H.C., Varley, J.M., Morgan, G.T. (1981). The transcription of satellite and ribosomal DNA sequences on lampbrush chromosomes of crested newts. In International Cell Biology (ed. H.G. Schweiger). Springer Verlag, Berlin, Heidelberg. pp. 33-46.

Macgregor, H.C., Sessions, S.K. and Arntzen, J.W. (1989). Evolutionary relationships among newts of the genus Triturus (family Salamandridae), from the viewpoints of comparative biochemistry, reproductive interactions and cytogenetics. J.Evol.Biol. in press.

Mancino, G., Ragghianti, M., Bucci-Innocenti, S. (1978). Experimental hybridisation within the genus Triturus (Urodela: Salamandridae). I. Spermatogenesis of F1 hybrids Triturus cristatus carnifex female x Triturus vulgaris meridionalis male. Chromosoma (Berl.). 69, 27-46.

Masumoto, H., Kenji, S., Tuneko, O. (1989). Alphoid satellite DNA is tightly associated with centromere antigens in human chromosomes throughout the cell cycle. Exp. Cell Res. 181, 181-196.

Mizuno, S., Macgregor, H.C. (1974). Chromosomes, DNA sequences and evolution in salamanders of the genus Plethodon. Chromosoma (Berl.) 48, 239-296.

Morgan, G.T. (1978). Absence of chiasmata from the heteromorphic region of chromosome 1 during spermatogenesis in Triturus cristatus carnifex. Chromosoma (Berl.) 66, 269-280.

Morgan, G.T., Macgregor, H.C., Colman, A. (1980). Multiple ribosomal gene sites revealed by in situ hybridisation of Xenopus rDNA to Triturus lampbrush chromosomes. Chromosoma (Berl.) 80, 309-330.

Neuer-Nitsche, B., Lu, X., Werner, D. (1986). Functional role of a highly repetitive DNA sequence in anchorage of the mouse genome. Nucleic Acid Res. 16.

Radic, M.Z., Lundgren, K., Hamkalo, B.A. (1987). Curvature of mouse satellite DNA and condensation of heterochromatin. Cell, 50, 1101-1108.

Sanchiz, F.B. (1977). Catalogo de los anfibios fosiles de Espana (Noviembre de 1977). Acta Geol.Hispan. 12, 103-107.

Sanchiz, F.B., Mlynarski, M. (1979). Pliocene salamandrids (Amphibia: Caudata) from Poland. Acta Zool. Cracov. 24, 175-188.

Sanchiz, F.B. and Szyndlar, Z. (1984). Filogeneza Triturus marmoratus. Spraw. Posiedz.Kom.Nauk, PAN, Krakow 26, 173.

Schmid, M., Olert, J., Klett, C.H. (1979). Chromosome banding in Amphibia III. Sex chromosomes in Triturus. Chromosoma (Berl.) 71, 29-55.

Sessions, S.K., Macgregor, H.C., Schmid, M., Haaf, T. (1988). Cytology, embryology, and evolution of the developmental arrest syndrome in newts of the genus Triturus (Caudata: Salamandridae). J.exp.Zool. 248, 321-334.

Sims, S.H., Macgregor, H.C., Pellatt, P.S., Horner, H.A. (1984). Chromosome 1 in crested and marbled newts (Triturus): An extraordinary case of heteromorphism and independent chromsome evolution. Chromosoma (Berl.) 89, 169-185.

Smith, G. (1979). Unequal crossing over and the evolution of multigene families. Cold Spring Harbor Symp. Quant. Biol. 38, 507-513.

Smith, G. (1976). Evolution of repeated DNA sequences by unequal crossing over. Science 191, 528-535.

Spurway, H., Callan, H.G., (1960). The vigour and male sterility of hybrids between the species Triturus vulgaris and T. helveticus. J. Genet. 57, 84-118.

Stephan, W. (1987). Quantitative variation and chromosomal location of satellite DNAs. Genet. Res. Camb. 50, 41-52.

Stephan, W. (1989). Tandem-repetitive noncoding DNA: forms and forces. Mol.Biol. Evol. 6, 198-212.

Varley, J.M., Macgregor, H.C., Erba, H.P. (1980a). Satellite DNA is transcribed on lampbrush chromosomes. Nature (Lond.) 283, 686-688.

Varley, J.M., Macgregor, H.C., Nardi, I., Andrews, C., Erba, H.P. (1980b). Cytological evidence of transcription of highly repeated DNA sequences during the lampbrush stage in Triturus cristatus carnifex. Chromosoma (Berl.). 80, 289-307.

Varley, J.M. and Macgregor, H.C. (1989). Interspecific homogeneity of a centromerically localised highly repeated DNA from Triturus (Amphibia, Urodela). Chromosoma (Berl.) in preparation.

Wallace, H. (1984). Chromosome 1 and sex determination of crested newts. Caryolgia 37, 3-8.

Wallace, H. (1987). Abortive development in the crested newt Triturus cristatus. Development 100, 65-72.

Wallis, G.P., Arntzen, J.W. (1988). Mitochrondrial DNA variation in the crested newt superspecies: limited cytoplasmic gene flow among species. Evolution 43, 88-104.

White, M.J.D. (1946). The spermatogenesis of hybrids between Triturus cristatus and T. marmoratus (Urodela). J. exp. Zool. 102, 179-207.

Wong, A.K.C., Rattner, J.B. (1988). Sequence organisation and cytological localisation of the minor satellite of mouse. Nucleic Acid Res. 16, 1645-1661.

Cytogenetics of Amphibians and Reptiles
Advances in Life Sciences

CYTOGENETICS AND CHROMOSOME SPECIATION IN EUROPEAN NEWTS OF THE GENUS *TRITURUS*

G. MANCINO

Dipartimento di Fisiologia e Biochimica, Laboratorio di Biologia cellulare e dello sviluppo, Universita' di Pisa, Pisa, Italy

SUMMARY: *Triturus* is a Palearctic genus which comprises 12 species and a still uncertain number of subspecies. The genus has been widely studied to define the biological characters, extensions and limits of the species, which are considered to be very conservative and to lie at the end of the process of morpho-anatomical and physiological differentiation. Hybridological results point out that the single newt species have reached different levels of evolution, judging from the different degree of efficiency and completedness of the reproductive isolating barriers. The species are also considered especially conformist as regards karyology: actually, the single chromosomes can undergo major rearrangements only sporadically, giving rise to chromosome polymorphisms that have a cohesive character and cannot be supposed to be involved in a short term chromosome speciation. The specific karyotypes have undergone a significant differentiation in amount and distribution of heterochromatin, as well as a long story of divergence and conservation of certain highly repeated DNA families. Established minute and large rearrangements, including some pericentric inversions usually considered neutral, can affect the genetic structure of the single rearranged populations and start a reproductive isolating barrier, as an incipient speciation mechanism. Therefore, chromosome speciation, should be considered as a long term evolutionary process.

INTRODUCTION

In line with the general theme of this Symposium, the present paper intends to discuss mainly some problems concerning the biological concept of the species in Triturus and the contributions given by experimental hybridization and cytogenetics of hybrids to investigate the origin, nature, and efficiency of the reproductive isolating mechanisms. Subsequently, the reports intends to mention the contribution given by a new approach to cytotaxonomy and cytogenetics provided by the hybridoma monoclonal antibody technique, which studies structure, organization, and function of protein components of lampbrush chromosomes (cf. Mancino, 1989), significantly improving the investigation on chromosome and genome factors which can play an important role in newt speciation. In fact, the fate of proteins associated with lampbrush chromosomes can be followed throughout embryogenesis and morphogenesis, and the information represent therefore a prerequisite for investigating any genetic variation in developmental protein pattern which may have important implications in the evolutionary diversification inside the genus Triturus. According to Larson (1980), in fact, the transformation of many morphological features can be achieved simultaneously by a genetic change that alters the growth rate of an organism such that extensive changes in shape result. Analogously, the rapid diversification of plethodontid salamanders is interpreted invoking a particular kind of heerochrony (that is the selective modulation of patterns of development) called paedomorphosis, by which the juvenile morphological features of an ancestor are retained in the adult morphology of a descendent (Wake, 1966). So, the link between modern cytogenetics and evolutionary biology can be emphatically considered an established achievement useful for a comprehensive view of the biological history of European newts.

CYTOGENETICS OF HYBRIDS AND THE BIOLOGICAL CONCEPT OF TRITURUS SPECIES

Triturus is a Palearctic genus including species and subspecies distributed from the Iberian Peninsula to Great Britain, from North to Central and South Europe, as far as Turkey, Syria, Lebanon and Israel. Paleontology has proved the antiquity of the genus: fossils belonging to *Triturus* species date back to the Belgian middle Oligocene and British upper Eocene (Sanchíz and Mlynarski, 1979; Estes, 1981; Milner et al., 1982). Fossils of *T. cristatus* and *T. marmoratus* have been found in the French and Spanish middle Miocene (Estes and Hoffstetter, 1976). Therefore, the present day species are supposed to have been differentiated since the first half of Miocene and the center for Salamandrid evolution, which probably began about 50 million years ago, was presumably Europe. The comparison of fossil findings with the present day distribution suggests that the family Salamandridae had a monophyletic Holarctic origin (Cei, 1962; Millner, 1983), even though some remnants from the Cretaceous of the Niger region may suggest a distribution of this family more extensive than thought (Estes, 1981). Anyway, the family Salamandridae was permanently established in Europe at least since the late Oligocene.

The number of extant *Triturus* species is still under discussion, mostly because properties, characters and limits of each species, as well as the taxonomic and evolutionary level of subspecies, are not defined yet. The number of species varies from 9 to 12 (as proposed by Bucci-Innocenti et al., 1983*a*; see also Frost, 1985; Mancino, 1988; 1989) and could increase if some of the numerous subspecies will be proved to have reached the status of real species. At present, the definition of species and subspecies is pursued by means of both hybridological studies and some advanced genetic and karyological techniques involving molecular biology of nucleic acids and biochemistry of proteins.

The hybridological approach to the problem of the species

Hybridological studies permit to evaluate the relatedness between morphologically different species and subspecies, the level of differentiation and position in the taxonomic hierarchy already reached by them, and the degree of perfection of their isolation barriers.

1. Spontaneous hybridization

Three cases of naturally occurring hybrids are known so far (Tab.I):

- the so-called T. blasii, arisen from the cross between T. cristatus and T. marmoratus in those areas of France where the two specific ranges overlap (cf. Vallée, 1959);
- the numerous hybrids arisen from the cross between T. vulgaris and T. montandoni in those areas of the Carpathian Mountains and Rumania, where the two species are sympatric and synthopic (cf. Ragghianti et al., 1978; Pecio and Rafiński, 1985);
- a case of a hybrid between T. vulgaris and T. helveticus recently detected in mid-Wales (Griffiths et al., 1987).

T. cristatus and T. marmoratus, differentiated as two distinct species since the late Neocene, still share many morpho-anatomical characters; relatively often, they share the breeeding places. When in syntopy, they are interbreeding species and give rise to the hybrid T. blasii, proving that the two parental species are not prevented from hybridization by an efficient reproductive ecological isolating barrier. However, Schoorl and Zuiderweijk (1981), Zuiderweijk (1986) and Zuiderwijk and Bouton (1987) point out that these related newt species are less syntopic than could be expected, although it is uncertain whether the occupation of separate habitats is due to competition, which generally occurs between hybridizing species. When the 2 species are syntopic, T. cristatus arrives earlier to the breeding pond (so it has a better chance to mate with the conspecifics) and uses the breeding period from the

Tab.I: Cases of spontaneous and experimental hybridization in the genus Triturus.

Hybridization	Method	Type	Cases
Spontaneous hybridization (in nature)*			T.marmoratus x T.cristatus T.v.vulgaris x T.montandoni T. v. vulgaris x T.helveticus
Experimental hybridization**	by interbreeding in captivity	between conspecific populations	always successful
		between conspecific subspecies	T.a.alpestris x T.a.apuanus T.v.vulgaris x T.v.meridionalis
		between closely related species	T.cristatus x T.carnifex T.cristatus x T.karelinii T.carnifex x T.karelinii T.v.meridionalis x T.montandoni T.v.meridionalis x T.helveticus T.v.meridionalis x T.italicus
		between species of undefined relatedness	T.v.meridionalis x T.vitt.ophryticus T.a.apuanus x T.vitt.ophryticus
	by in vitro insemination	between apart species	T.v.meridionalis x T.a.apuanus T.carnifex x T.v.meridionalis T.carnifex x T.italicus T.carnifex x T.a.apuanus
		between different genera	Triturus x Salamandrina Triturus x Notophthalmus

* cf. Vallée(1959), Ragghianti et al. (1978), Mancino et al. (1979a), Griffiths et al. (1987).

** cf. Lantz (1947), Callan and Spurway (1951), Lantz and Callan (1954), Mancino and Scali (1964), Mancino (1968, 1988), Mancino et al. (1978), Mancino et al. (1979a, b).

beginning to the end, whereas T. marmoratus arrives only later. Heterospecific matings probably take place between late immigrating T. cristatus and early arriving T. marmoratus specimens. Zuiderweijk and Bouton (1987) suppose that interbreeding leads to a local extinction of these species, which affects more severely T. cristatus than T. marmoratus. This may be an example of direct extinction by interspecific competition, indicated as reproductive selfdestruction by Zaslavsky (1963, 1967). According to White (1978), this mechanism is "able to determine the stability of the contact zones and to make it impossible for allopatric species to be distributed more widely. ... The descendants of immigrants from the overlapping zone into a pure population of another species are doomed to extinction".

Interbreeding between T. vulgaris and T. montandoni is only occasional and the hybrid zone very narrow; however, the occurrence of spontaneous hybrids indicates that the two species are not prevented against hybridization by efficient eco- and ethological premating isolation mechanisms. In particular, ethological isolation seems to be scarcely efficient, although newt species release a series of species-specific tactile, olfactory and broadly behavioural stimuli which lie at the root of premating isolating mechanisms. There is evidence in newts that sexual persuasion of the female is accomplished by male glands that are employed during the intermediate stage of courtship. Most salamanders are nocturnal and, not surprisingly, many aspects of their sexual behaviour are chemically mediated: the functions of chemical signals include mate attraction, orientation and coordination during sperm transfer, and sexual persuasion (Arnold and Houck, 1982). Male newts can identify conspecific females even by their skin secretions alone (Twitty, 1955; 1961), but female choice should be the major factor determining species isolation in newts, since males sometime display towards both other males and females of different species, and to inanimate objects as well (Malacarne and Giacoma, 1986). Anyway, despite the highly species-specific sequence of stimuli and reactions (Halliday, 1975, 1977; Pecio and Rafinski, 1985; Sparreboom and Arntzen, 1987), the two species interbreed.

Two other closely related species, T. vulgaris and T. helveticus, largely share seasonal microhabitat and feeding niche dimensions and are said to have developed species-specific ethological isolating barriers (Halliday, 1975; 1977), before converging to utilize common resources (Griffiths, 1987). However, the efficiency of such ethological barriers is not fully accomplished according to the results obtained by Mancino and Scali (1964) and the recent detection of natural hybrids by Griffiths et al. (1987). Therefore, the behavioural pattern of courtship does not represent a reliable and effective premating mechanism to prevent hybridization. The examples of naturally occurring hybrids demonstrate also that the inner premating barrier, involving the cell level i.e., gametic incompatibility, is only partially efficient.

Thus, prevention against hybridization seems to be due to a cooperation of premating and postmating factors, acting sequentially and progressively: when the premating barriers fail, the postmating mechanisms operate, disrupting any attempt of hybrid viability by means of reduction of reproductive success, early blockage of embryogenesis, lethal abnormalities in morphogenesis, and, eventually, meiotic anomalies, sterility and final failure of possible F1 hybrids. In any case, speciation cannot occur via hybridization: established hybrid lineages do not exist in nature and reports concerning a phenomenon of hemiclonal reproduction sensu Kalman (in Vrijnhoek et al., 1977) or hybridogenesis (Tunner, 1974) in a hybrid population of "T. vulgaris" of Lake Cracurele, in Rumania (Fuhn et al., 1973) need further genetic and cytogenetic evidence, such as that recently obtained in the hybridogenetic water frogs belonging to the so-called "Rana esculenta" complex (Hotz et al., 1985).

2. Experimental hybridization

The taxonomic and phylogenetic relationships between newt species, as well as the degree of specific and subspecific differentiation, can be experimentally tested by using two sequential hybridization methods:

a) natural hybridization in captivity (interracial and interspecific matings obtained under laboratory conditions);

b) experimental hybridization (*in vitro* insemination of eggs with heterospecific sperm), when method a fails.

Method a gives results proportional to the degree of morphological relatedness between the two parents: the greater is the extent of affinities, the higher is the rate of fertilization, development and metamorphosis of hybrids. Both the regularity of meiosis and the fitness are nearly normal in F1 hybrids between conspecific populations (which always successfully interbreed, Tab.I), and progressively decrease in F1 hybrids arisen from crosses between either conspecific subspecies (which easily interbreed, as the two examples reported in Tab.I clearly show) or akin species. According to this rule, *T. a. alpestris* and *T. a. apuanus* still represent two distinct subspecies. As regards *T. a. cyreni*, *T. a. reiseri*, *T. a. serdarus*, *T. a. veluchiensis*, which show a clear chromosome differentiation in heterochromatin distribution (Herrero and Arano, 1985; 1987), they cannot be considered as true species, having not been tested with method a yet. The same test should be performed to define the exact rank of *T. a. inexpectatus*, recently described by Dubois and Breuil (1983), as well as *T. a. montenegrinus* and any other supposed subspecies of the Balkan area (cf. Thorn, 1969). On the other hand, by evaluating all the hybridological results obtained with method a and analyzing karyological and cytogenetic findings (Callan and Spurway, 1951; Spurway, 1953; Lantz and Callan, 1954; Mancino et al., 1977 and not published), Bucci-Innocenti et al. (1983a) and Mancino (1988; 1989) have suggested that *T. c. cristatus*, *T. c. carnifex*,

T.c.dobrogicus and *T. c. karelinii* must be regarded as true closely related species, rather than geographical races of *T. cristatus*. Moreover, results obtained with method *a*, integrated by cytogenetic data (Mancino, 1959; Mancino and Scali, 1964), point out that *T. vulgaris*, *T. montandoni* and *T. helveticus*, which interbreed in nature and easily in captivity, are closely related species. The relation of *T. vittatus* is uncertain: in fact, it mates spontaneously, although in captivity, with both *T. v. meridionalis* (belonging to the so-called *Paleotriton* subgenus, according to Bolkay, 1928) and *T. a. apuanus* (*Mesotriton*), but never interbreeds with *T. carnifex* (*Neotriton*).

A final consideration should be done as regards method *a*: when mating under laboratory condition is successful, captivity seems to play an important role, removing eco-, chrono-, and ethological barriers and harmonizing the sexual behaviour of both parents.

Method *b* i.e., *in vitro* insemination, is used when ecological, temporal and ethological premating isolation barriers against hybridization are completed, being the 2 parental species distantly related. An example is given by *T. v. meridionalis* and *T. carnifex*, two ecologically and ethologically well segregated species which never interbreed in nature. They are sympatric over almost the whole of their distributional areas, and are similar in their preference for breeding ponds within which they seem to occupy different ecological niches as to avoid the effects of the so-called "competitive exclusion principle" (cf. Dolmen, 1988). In addition, *T. carnifex* seems to arrive to the pond earlier and *T.vulgaris* later on and possibly to exhibit a resource partitioning, although the habitat requirements are much the same. Yet, method *b* permits to evidentiate a residual gametic compatibility between their gametes, which can still attract each other and interact. Similarly, some more distantly related newt species display a residual gametic compatibility, when their gametes are submitted to *in vitro* insemination (Tab.I).

A certain potential for hybridization persists even between _Triturus_ and other Salamandrid genera distinguished by genus-specific peculiarities (Wake and Özeti, 1969), among which behavioural traits represent an unsurmountable ethological barrier both in nature and in laboratory following method _a_. This potential, albeit generally very low, is a little higher when genera are taxonomically not very distant i.e., _Tylototriton_ and _Pleurodeles_, which share similar karyotypes and genomes (Ferrier and Beetschen, 1973). When genera are more deeply segregated i.e., _Triturus_ and _Salamandrina_ (both sharing the same chromosome number: 2n=24 n=12) or _Triturus_ and _Notophthalmus_ (divided also by a different chromosome number, which is 2n=22 n=11 in the New World newt), gametes show little, if any capacity of attracting each other. In all these instances, anyhow, the development of possible experimental intergeneric F1 hybrids is precociously blocked, owing to qualitatively and quantitatively unbalanced mitoses (as in _Triturus_ x _Notophthalmus_) and alterations in gene dosage and gene expression (as in _Triturus_ x _Salamandrina_). In the latter case, an exterior manifestation of a deeply altered gene dosage is represented by the morphogenesis of the hind limb, since hybrid larvae die when gene dosage has become maximally altered and they are not able to develop a hind limb having either 5 fingers, as in _Triturus_, or 4, as in _Salamandrina_. Certainly, all organs and functions of the intergeneric larvae are affected by an irregular control of gene expression, since hybrid larvae have to overcome profound anatomical and physiological changes at metamorphosis, such as, for instance, the development of a respiratory system based on the choice of either the presence (_Triturus_) or the absence (_Salamandrina_) of the lungs.

Results are completely negative when a cross between different families is attempted. For instance, not even a partial egg activation is attained after _in vitro_ insemination in both crosses between _Triturus_ (Salamandridae) and _Hydromantes_ (Plethodontidae).

These findings point out that salamandrid species and genera can still evolve, but only within the limits of the family, which therefore represents a natural rather than a conventional taxon.

In conclusion, genera and species of Salamandridae have reached different levels of evolution, judging from the different degree of efficiency and completeness of the reproductive isolating barriers shown by the single taxa during the hybridological tests. It can be argued that, during evolution, the potential for hybridization has slowly decreased in inverse ratio to the efficiency of isolation barriers. It can also be inferred that, within _Triturus_, the morpho-anatomical and physiological differentiations might have been faster than the evolution of reproductive isolating mechanisms. These processes of specialization appear still in progress and can involve, in a different rate, single organs and functions in closely related species.

Cytogenetic and genetic evidence for evolutionary systematics

Within the framework of the hybridological, cytotaxonomic and cytogenetic results available so far, some caution is advised before accepting at its face value the rigorous division of the genus _Triturus_ into 3 subgenera (Bolkay, 1928), more or less equivalent to the 3 groups of Lantz (1947). However, two trends of karyological evolution are evident within the genus: the first one is shared by _T. cristatus_ and its 4 strictly allied species; the second one involves all the remaining newts, including _T. vittatus_ (Fig. 1).

The karyotypes of _T. cristatus_ and its 4 akin species derive from a common ancestral karyotype, as proved by the particular structure of long arms of chromosomes I which, in both sexes and in all the 5 allied species, are heterozygous for length value and largely heterochromatic, showing an arm- and species-specific distribution of dark C-bands (Mancino et al., 1973, 1977; Rudak and Callan, 1976; Morgan, 1978; Wallace,

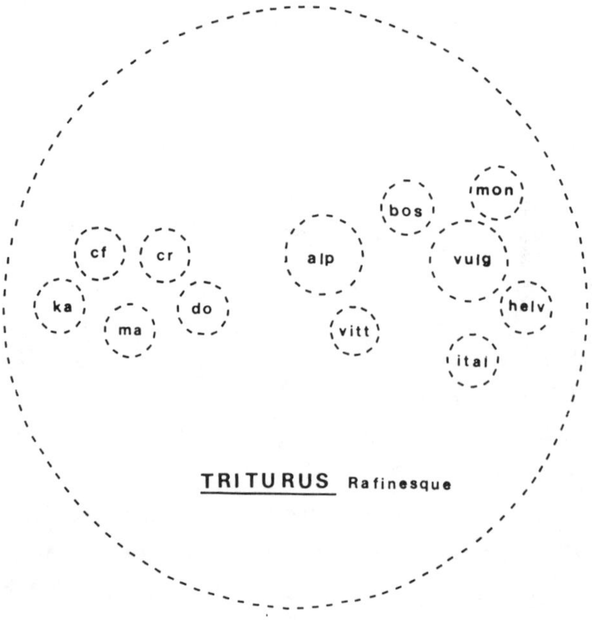

Schematic representation of the genus Triturus, based on hybridological and cytogenetic results. The genus is not subdivided into subgenera; however, two evolutionary trends are clearly visible, which may correspond to two species groups or superspecies.

1984). The corresponding lampbrush bivalent I is asymmetric, owing to the different length of the two long arms and the heterozygous distribution of some cytogenetic markers i.e., the contorted loops (Callan and Lloyd, 1960; Nardi et al., 1972). The long arms are asynaptic and achiasmatic in both female (Fig.2) and male meiosis. The heterozygosity for length and for different distribution of both C-bands on mitotic chromosomes I and contorted loops on the corresponding lampbrush chromosomes is a prerequisite for a viable development, since all embryos homomorphic for chromosomes I arrest and die at late tailbud stage (Macgregor and Horner, 1980); even hybrid embryos between <u>T. marmoratus</u> and <u>T. cristatus</u> seem to show the same syndrome as to prove that this particular chromosome

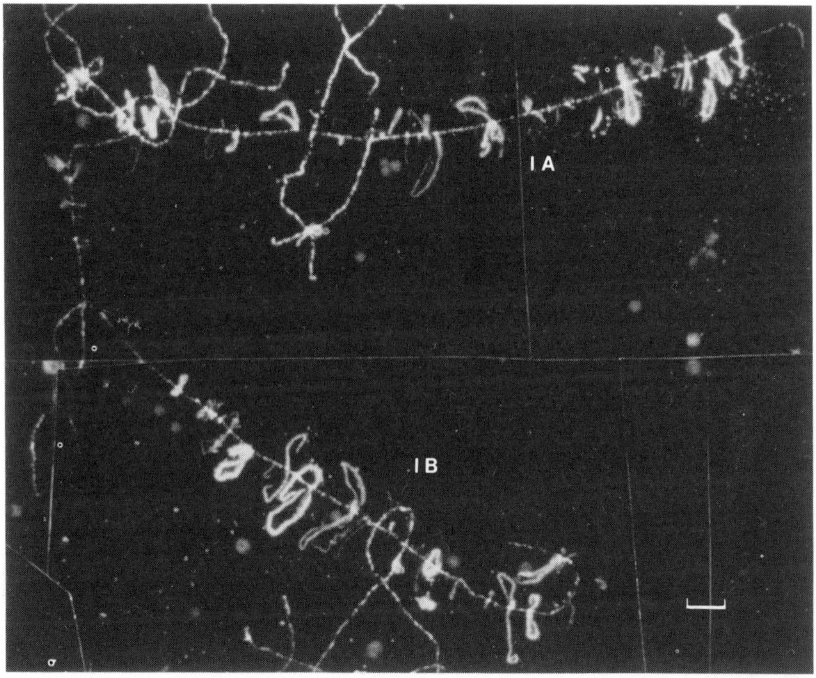

Fig.2: Lampbrush bivalent I of T. marmoratus immunostained by mAb A1/5 (1:750). The picture results from two overlapping exposures, the first revealing the chromosome axis by DAPI (4′-6-diamidino-2-phenylindole) staining, the second revealing rhodamine-fluorescing loops. IA = The longer partner; IB = the shorter one. Bar represents 20um.

character was present prior to the divergence of the two species (Macgregor, 1982). Macgregor (1982) and Macgregor and Sessions (1986a) report that the long arm of chromosome I in T. cristatus represents 7.9% of the genome and the heteromorphic region of that arm comprises 5.8% of the genome. In absolute terms, this means that the heteromorphic region contains 1.33 pg of DNA or 1.2 x 10 raised to 9th power base-pairs. The heteromorphic region of chromosome I is disproportionately rich in highly repetitive DNA, most of which belongs to a few families of satellite DNA sequences that together constitute between 3 and 4% of the genome (Varley et al., 1980).

The view of the genus proposed in Fig.1 is approximate, because it does not take into account either the interspecific genetic distances or the rate of derivation of the single species; moreover, it does not report all the subspecies morphologically or enzymatically recognized so far. Several species have been intensively studied by means of electrophoretic analysis of proteins and often the existence of sibling subspecies has been revealed only enzymatically. It can occur, in fact, that the degree of genetic differentiation is not in accordance with the pattern of morphological diversity. For instance, T. v. dalmaticus should be considered as a true subspecies, because, albeit its high morphological similarity to T.v.graecus, is deeply genetically differentiated from it (Kalezic', 1984). Within T. alpestris, T.a.reiseri, being found genetically mostly similar to T.a.alpestris, has not been revalidated as a subspecies, while the subspecific status of T.a. veluchiensis, considered doubtful up to now, seems fully proved (Arano and Arntzen, 1987). The electrophoretic analysis of proteins has also evidentiated that, in T. italicus, the intraspecific populations, morphologically very similar to each other, appears genetically differentiated, although not to such an extent as to be considered subspecies (Ragghianti and Wake, 1986).

Extensive studies on the genetic structure of the single populations, such as those performed by Larson (1980) in Aneides flavipunctatus, Kalezic' and Tucic' (1984) in T. vulgaris and, above all, Ragghianti and Wake (1986) in T. italicus, represent a correct attempt to monitor genetic changes at structural gene loci in several populations distributed all over the range of the species. The significant genetic variations evidentiated by this research suggest that the evaluation of intraspecific differentiations should be always preliminary to that of the interspecific ones, in order to obtain more reliable values of the genetic distance between species. This assumption seems to be in line with Wallis and Arntzen (1989), who found that the mitocondrial DNAs of T. cristatus, T. carnifex, T. dobrogicus and T. karelinii are clearly differentiated as those of real species, and were also able to correlate mtDNA variation with the morphological features defining the single species. Moreover, Wallis and Arntzen (1989) have reported that T. cristatus and T. dobrogicus are genetically homogeneous over most or all of their ranges, while T. carnifex and T. karelinii are notably differentiated as far as mitocondrial DNA is concerned: there are some populations of T. carnifex and T. karelinii which seem to be genetically as much differentiated as semispecies or sibling species. According to this intraspecific differentiation, the crested newts may comprise a complex taxonomic hierarchy of semispecies and sibling species, the exact level of which must be carefully investigated, suggesting an appropriate integration of the results of enzyme analysis with morpho-physiological, eco-ethological, hybridological and cytogenetic criteria, before evaluating the intra- and inter-specific distances and hazarding any hypothesis on the possible modes of speciation.

CHROMOSOME SPECIATION AND EVOLUTION

As a rule, the efficiency of reproductive isolating barriers varies to an extent depending on the degree of the taxonomic relatedness existing between the parents: it is higher between two species belonging each to a different evolutionary trend occurring in the genus, lower between species belonging to the same trend (Fig. 1). In other words, the efficiency of the reproductive isolation is related to the differentiation of the specific karyotypes, namely the cytotaxonomic relatedness

(Mancino et al., 1977; Bucci-Innocenti et al., 1983a).

The present day newts are considered especially conformist: in fact, they share the chromosome number, the number of chromosome arms, as well as the general shape of the karyotype, that appears settled into a remarkably stable pattern of metacentrics and submetacentrics. Yet, the single chromosomes have undergone a significant differentiation in amount and distribution of heterochromatin, as banded karyotypes clearly show (Mancino et al., 1977; Bucci-Innocenti et al., 1983a).

At present, only sporadically and accidentally whole chromosomes undergo major rearrangements, that give rise to structural heterozygotes. Such heterozygotes, encountering meiotic difficulties affecting fertility, are eliminated by natural selection. Possible changes cannot become fixed by means of genetic drift or meiotic drive and spread on a part of the specific range. Hereby, as a rule, there exist no intraspecific variability and the newt species are not divided into differentiated cytotypes and biotypes (Mancino et al., 1970; Hartley and Callan, 1977; Mancino, 1988).

An exception is represented by a multiple chromosome polymorphism, widespread over nearly all the specific range and due to 2 pericentric inversions and a reduction of the pericentromeric heterochromatic C-band on long arms of chromosome XII of *T. italicus*. Several karyomorphs are present, all showing practically normal meiosis, since the restructured part occurs in a section which is void of chiasmata (Ragghianti et al., 1980; Bucci-Innocenti et al., 1983b). Heterozygotes are somehow advantageous and retain a high rate of fertility, which, according to Hedrick (1981), may be responsible for the diffusion and spreading of the rearrangements over the range and their fixation at least in small and geographically isolated populations, which maintain a high rate of interbreeding with the primitive ones and therefore cannot be grouped into differentiated cytotypes. Chromosome polymorphisms, wherever detected, have a cohesive character and cannot be supposed to be involved in a short-term chromosome speciation, but they can be interpreted as a transitional step from chromosome to karyotype evolution. Accordingly, it is conceivable to suppose that the few heterobrachial chromosomes still enduring in all salamandrid karyotypes are instable and should be rearranged by following, under heavy selection control, the same sequence repeatedly pursued inside the family in its evolutionary history (White, 1973), in order to establish the stable pattern of the basic karyotype.

The most remarkable case of chromosomal instability is represented by the well known chromosome I of _T. cristatus_ and allied species, whose heteromorphic region is highly variable as regards C-bands, location of specific gene sequences (i.e., distribution of ribosomal genes), lampbrush loop landmarks revealed by means of both traditional methods (Callan and Lloyd, 1960) and the hybridoma monoclonal antibody technique (Ragghianti et al., 1988; Mancino, 1988; 1989) (Fig.2). Such variability has been found between and within species, and even between individuals of the same population (Macgregor and Sessions, 1986_b_). Sims et al. (1984) and Macgregor and Sessions (1986_b_) have speculatively suggested that the evolution of chromosome I from an ancestral karyotype of the present day _T. cristatus_ and allied species could have been characterized by a series of minor and major rearrangements leading to the present karyotypes.

If a relatively short term involvement of a chromosomal change in speciation cannot be proved, more caution is required before denying possible long-term effects of minute and large rearrangements on the genetic structure of single populations, possibly increasing the mutation rate of the gene pool and favouring the genetic differentiation as to give rise to an isolation mechanism between the original population and the restructured one. Then, an established rearrangement which presumably lacks a short term involvement in speciation, may have a long term evolutionary effect. A common change seems to be represented by quantitative variation in centromeric heterochromatin and satellite DNA which are supposed to be correlated to chromosome behaviour at mitosis, rate of cell proliferation in organismal development, gene activity: natural selection, for instance, may favour those changes in satellite DNA related to newt ecology, which reduce cell cycle times when rapid rates of growth are advantageous at some stages of development.

This suggests that molecular evolution and evolution of chromosomes are not disconnected from evolution of the whole species, even if the modes of the interconnection are not apparent and immediately comprehensible. On this regard, the reconstruction of the evolutionary molecular events of highly repeated DNA families (origin, divergence and conservation of satellite DNA), which represent a large part of the genome, might reveal important aspects of genome and karyotype evolution and, possibly, the relationships existing between molecular evolution and past history of the species. In this contest, it is worth while mentioning a model proposed by Macgregor and Sessions (1986 a and b) on the basis of the patterns of variation of heterochromatin and cloned satellites among newts, intending to explain the localized growth and subsequent dispersal of heterochromatin and satellite DNA in single chromosomes. Satellite DNA sequences would arise at centromeres, where they would tend to accumulate (as centromeric C-heterochromatin), mainly by repeated chromosomal exchanges or tandem duplications. Subsequently, small chromosomal rearrangements (such as inversions and unequal exchanges) would occur at random near the centromere, dispersing satellite DNA and producing pericentric C-bands at the expense of centromeric C-bands. Further chromosome rearrangements would lead to the progressive dispersal of satellite DNA and heterochromatin away from the pericentric regions throughout the genome, along chromosome arms up to telomeres. Therefore, in newts, centromeric families would be of "young" phylogenetic age, pericentromeric families of "intermediate" age and the dispersed ones would represent ancient or "old" families (Macgregor and

Sessions, 1986 a). The model could explain both the patterns of variation in heterochromatin and satellite DNA among extant newt species and the large changes in genome size that involve the proportional growth of chromosomes without producing any change in chromosome number or shape. The waves of dispersal of satellite DNA from centromeres to telomeres are supposed to happen again and again in rapid succession, one wave overlapping another, by means of many rearrangements of different sort. Also an alternative explanation, recently proposed by Cremisi et al. (1988 a and b), giving a dynamic image of newt chromosomes within the frame of an apparently static karyotype, supports the opinion that the "evolutionary life" inside the single chromosomes of the specific genomes is still very active: then, each chromosomal event may possibly enhance the mutation rate of the single populations and cause a long term change in their genetic structure, eventually leading to a subspecies differentiation by starting an interpopulation reproductive isolation showing a gradually increasing efficiency.

Also the discovery of several stages of morphological differentiation of sex chromosomes has permitted the determination of given evolutionary series of karyotypes, in which the increasing structural complexity of sex chromosomes can be reconstructed (Schmid et al., 1979; 1988). The rate of sex heterochromatinization seems to vary in different newt species; moreover, the process involves a chromosome pair which can vary inside the genus, being shared (pair IV) only by the karyotypes of T. cristatus and allied species. Inside this group of related species, the differentiation of sex chromosomes must have been very old and it does not seem to have contributed to start a short-term chromosome speciation process. The same assumption can be done for the species belonging to the evolutionary trend of T. vulgaris, whose karyotypes do not share location and rate of sex heterochromatinization (pairs II to IV). These findings are believed to support the presence of two karyological and evolutionary trends distinguishing the genus Triturus as shown in Fig.1.

Cytological differentiation of sex chromosomes is often said to "accompany" speciation processes, rather than to induce them. However, the differential sex heterochromatinization appears as an autonomous, although severely controlled, evolutionary property of the karyotypes, which could contribute to change the genetic structure of populations and to start a novel reproductive isolating mechanism eventually leading to speciation.

CONCLUSIONS

The importance of experimental hybridization and cytogenetics of hybrids lies in their possibility to offer to systematics a biological approach, which can greatly contribute to encompasse the characteristics, genetic status, developmental properties, and evolutionary history of the single entity, the species, in nature. Above all, they represent a reliable means in defining the boundaries and contours of the populations, since some newt species are divided into populations connected by reproductive continuity and presenting a varying level of introgression, some others are represented by partially reproductively isolated populations. In particular, the results underline the necessity to improve the study of organism and population diversity and intraspecific genetic structure, before drawing any deductions on possible interspecific genetic distances. The findings are also in line with cytotaxonomic deductions and support the opinion that the genus has a cohesive character and that there is no need to subdivide it into subgenera, although two evolutionary trends can be recognized within it hybridologically and cytogenetically. Finally, we can assume that speciation in newts cannot occur via hybridization, since meiosis of F1 hybrids lacks cytological mechanismms for regularization: spermatogenesis of diploid hybrid specimens is disturbed to an extent depending on the cytotaxonomic relatedness between parents, complete asynapsis representing the most serious meiotic anomaly among spermatocytes I (Fig.3). Instead, meiosis

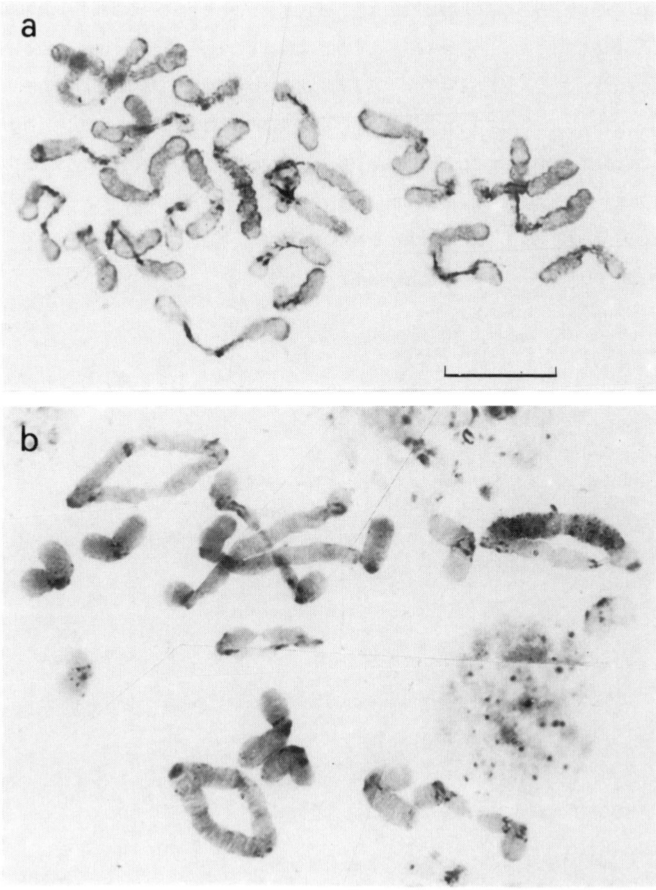

Fig.3: Primary spermatocytes at metaphase I of a hybrid arisen from (a) the cross between two distantly related species (T.carnifex x T.v.meridionalis) and (b) two closely related species (T.carnifex x T.marmoratus). a = asynapsis is total; b = pairing and chiasma formation occur between homologous, although heterospecific chromosomes; in this hybrid, synapsis can be complete and number of chiasmata close to that of the parental species.

of F1 diploid hybrid female specimens seems to attempt a sort of regularization by avoiding pairing and chiasma formation, which could complicate the possible elimination of the foreign genome and the reconstitution of a qualitatively and quantitatively balanced genome: asynapsis is therefore the commonest figure of oocytes at meiosis I, even from diploid hybrids between closely related species (Fig.4). Allopolyploid specimens cannot be regularized by combining a polyploidization with hybridization and the formation of chiasmata between homospecific chromosomes can lead to a more exit of repartition at meta-anaphase I than complete asynapsis, and eventually to the final failure of gametogenesis.

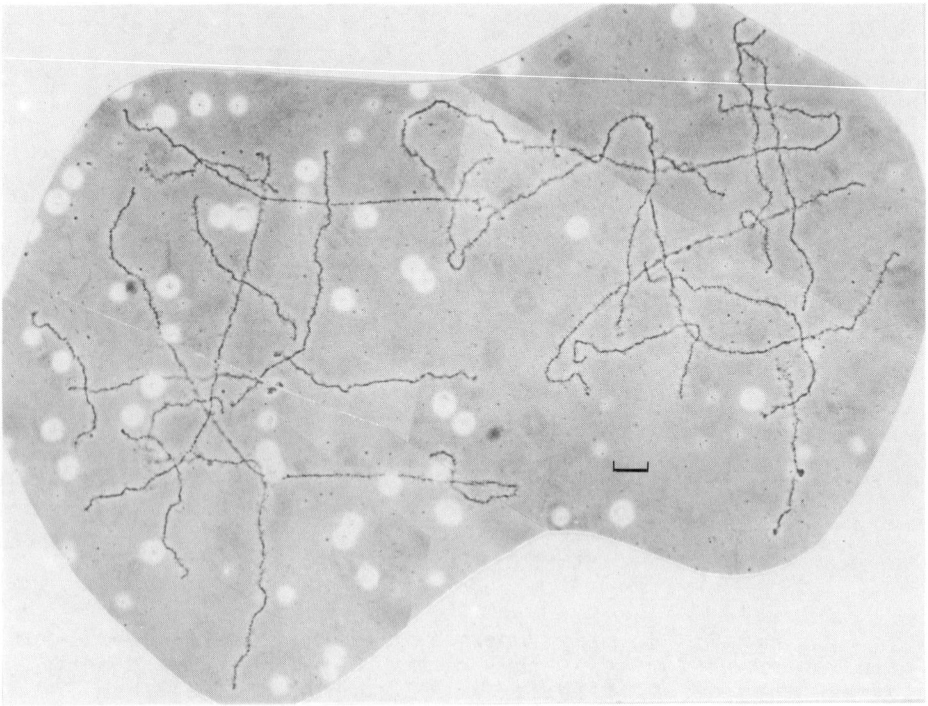

Fig.4: Unpaired lampbrush chromosomes of an asynaptic oocyte from a diploid <u>T. blasii</u> female; some of the univalents are joined by end-to-end heterologous fusions.

Cytogenetics has provided a body of results which indicate that occasional chromosome changes can still occur in the present day newt species, but changes of single chromosomes can be tolerated only if they follow the pattern already experienced by most chromosomes of the complement, in order to reach the stable pattern of the specific karyotypes. Therefore, chromosomes can change within the framework of a strict karyotypic orthoselection, as a step of a more general karyotype evolution: this sort of changes must be somewhat advantageous, as they quickly spread out giving rise to chromosome polymorphisms, by which they become established. Instead, there is no evidence of a chromosomal change related to a short term chromosome speciation in newts, since there are still many unknown variables concerning the factors responsible for the fixation of novel chromosomal rearrangements in natural populations and their role in evolution. No cases in newts can be referred to the well known example of chromosome speciation, called stasipatric speciation (White, 1978), since rearrangements are generally disadvantageous, inducing serious meiotic disturbancies, and are soon eliminated. Likewise, chromosomal rearrangements in newts do not fit to any other models of chromosome speciation, such as the cascading or chain process model proposed by Hall (1983) to explain the high species diversity and chromosomal variability in Sceloporus: in spite of the fact that White (1978) considered Sceloporus as examplifying chromosomal speciation in progress, the cascading chromosomal speciation hypothesis appears unlikely, according to the results of a computer simulation study performed by Sites et al. (1988) within the constrains of the model. However, any chromosomal change cannot be excluded to affect the genetic structure of a restructured population in a long term, so determining incipient reproductive isolating mechanisms involving a subsequent speciation.

At present, cytogenetics in newts appears as a renewed and a still promising field of investigation, since chromosome and karyotype studies can rely on both increasing molecular biology of nucleic acids and the use of monoclonal antibodies on giant lampbrush chromosomes (Ragghianti et al., 1988; Mancino, 1988; 1989). The latter methodology, in particular, portends exciting developments of cytotaxonomy and cytogenetics in the near future, not only because it ensures the availability of detailed functional cytogenetic maps, but also because the fate of certain chromosomal proteins can be followed throughout embryogenesis, as well as their role in transcription at and after the lampbrush phase, in nuclear determination, cell differentiation, tissue interaction, and during some morphogenetic processes such as metamorphosis. Cytogenetics, therefore, is going to play a novel role in phylogenetic systematics and evolutionary biology, since it can use a variety of biomolecular, biochemical, and other sophisticated approaches to the problem of chromosome speciation, which is essential for the comprehension of the real evolutionary potentialities of the species and the basic processes of life.

Acknowledgments: I am grateful to Matilde Ragghianti and Stefania Bucci for criticism and kind assistance in preparing the manuscript.

REFERENCE

Arano, B., and Arntzen, J.W. (1987) 4th Gen. Ord. Meet. S.E.H. Nijmegen (van Gelder, J.J., Strijbosch, H., and P.J.M. Bergers, Eds), pp.21-24.
Arnold, S., and Houch, L. (1982) In: Biochemical aspects of evolutionary biology (M. Nitecki, Ed), Univ. Chicago Press, pp.173-211.
Bolkay, St.J. (1928) Z. Anat. Entw. Gesch. 86, 259-319.
Bucci-Innocenti, S., Ragghianti, M., and Mancino, G. (1983a) Copela 1983 (3), 662-672.
Bucci-Innocenti, S., Ragghianti, M., and Mancino, G. (1983b) Chromosoma (Berl.) 88, 208-215.
Callan, H.G., and Lloyd, L. (1960) Phil. Trans. Roy. Soc. London B 247, 135-219.
Callan, H.G., and Spurway, H. (1951) J. Genet. 50, 235-249.
Cei, J.M. (1962) Scientia (Bologna) 56(6), 1-58.
Cremisi, F., Vignali, R., Batistoni, R., and Barsacchi, G. (1988a) Chromosoma (Berl.) 97, 204-211.
Cremisi, F., Vignali, R., Batistoni, R., and Barsacchi, G. (1988b) Cytotecnology 1, 185-188.
Dolmen, D. (1988) Amphibia-Reptilia 9, 365-374.
Dubois, A., and Breuil, M. (1983) Alytes 2, 9-18.
Estes, R. (1981) In: Handbuch der Paleoherpetology (Wellnhofer, P., Ed) G. Fischer Verlag, Stuttgart part 2, pp. 1-115.
Estes, R., and Hoffstetter, R. (1976) Bull. Mus. Nat. Hist. Paris, ser. 3, 398, 297-343.
Ferrier, V., and Beetschen, J.-C. (1973) Chromosoma (Berl.) 42, 57-69.
Frost, D.R. (1985) Amphibian species of the world. A taxonomic and geographical reference. Allen Press Inc. & The Assoc. Syst. Coll., Lawrence Kansas, pp. 1-732.
Fuhn, E., Şova, M., and Dumitrescu, M. (1975) Museul de Stiintele Naturii, Studii si Comunicari, pp. 225-236.
Giacoma, C., and Balletto, E. (1988) Boll. Zool. 55, 337-360.
Griffiths, R.A. (1987 J. Anim. Ecol.56, 441-451.
Griffiths, R.A., Roberts, J.M., and Sims, S. (1987) J. Zool. London 213, 133-140.
Hall, W.P. (1983) In: Advances in herpetology and evolutionary biology (Rhodin, A.E.J., and K. Miyata, Eds), Museum Comp. Zool., Cambridge, Mass., pp.643-679.
Halliday, T.R. (1975) Zool. J. Linn. Soc. 56, 291-300.
Halliday, T.R. (1977) In: The reproductive biology of Amphibians (Taylor, D., and Guttman, S.I., Eds), Plenum Press, New York, pp.185-232.
Hartley, S.E., and Callan, H.G. (1977) J. Cell Sci. 24, 131-141.
Hedrick, P.W. (1981) Evolution 35, 322-332.
Hotz, H., Mancino, G., Bucci-Innocenti, S., Ragghianti, M., Berger, L., and Uzzell, T. (1985) J. Exptl. Zool. 236, 199-210.

Kalezić, M.L. (1984) Amphibia-Reptilia 5, 221-230.
Kalezić, M.L., and Tucić, N. (1984) Evolution 38, 389-401.
Lantz, L.A. (1947) Latr. Proc. Zool. Soc. London 117, 247-258.
Larson, A. (1980) Evolution 34, 1-17.
Lantz, L.A., and Callan, H.G. (1954) J. Genet. 52, 165-185.
Macgregor, H.C. (1982) In: Genome evolution (Dover, G.A., and R.B. Flavell, Eds), Academic Press, London, pp. 325-341.
Macgregor, H.C., and Horner, H. (1980) Chromosoma (Berl.) 76, 111:122.
Macgregor, H.C., and Sessions, S.K. (1986a) Phil. trans. Roy. Soc. London B 312, 243-259.
Macgregor, H.C., and Sessions, S.K. (1986b) Verh. Dtsch. Zool. Ges. 79, 137-148.
Malacarne, G., and Giacoma, C. (1986) Boll. Zool. 53, 79-83.
Mancino, G. (1959) Atti Soc. Tosc. Sci. Nat. Mem. B 65, 70-90.
Mancino, G. (1968) Atti Acc. Naz. Lincei 44, 697-700.
Mancino, G. (1988) In: Il problema biologico della specie (Ghiara, G., Luporini, P., Mancino, G., and R. Nobili, Eds), Collana U.Z.I.: Problemi di biologia e di storia della natura, Mucchi Editore, Modena 1, pp.211-237.
Mancino, G. (in press) In: Amphibian cytogenetics and evolution (Green, D., and S.K. Sessions, Eds), Academic Press New York.
Mancino, G., Nardi, I., and Barsacchi, G. (1970) Cytogenetics 9, 260-271.
Mancino, G., Ragghianti, M., and Bucci-Innocenti, S. (1973) Atti Acc. Naz. Lincei 55, 559-564.
Mancino, G., Ragghianti, M., and Bucci-Innocenti, S. (1978) Chromosoma (Berl.) 69, 27-46.
Mancino, G., Ragghianti, M., and Bucci-Innocenti, S. (1979a) Caryologia (Firenze) 32, 61-79.
Mancino, G., Ragghianti, M., and Bucci-Innocenti, S. (1979b) Chromosoma (Berl.) 73, 207-226.
Mancino, G., Ragghianti, M., and Bucci-Innocenti (1977) In: The reproductive biology of Amphibians (Taylor, D.H., and Guttman, S.I., Eds), Plenum Press, New York, pp. 411-447.
Mancino, G., and Scali, V. (1964) Caryologia (Firenze) 17, 161-194.
Morgan, G.T. (1978) Chromosoma (Berl.) 66, 269-280.
Milner, A.R. (1983) In: Evolution: time and Space. The emergence of the Biosphere (Sims, R.W., Price, J.H., and P.E.S. Walley, Eds), Academic Press, London, pp. 431-468.
Milner, A.C., Milner, A.R., and Estes, R. (1982) Tertiary Res. 4, 149-154.
Nardi, I., Ragghianti, M., and Mancino, G. (1972) Chromosoma (Berl.) 40, 321-331.
Pecio, A., and Rafiński, J. (1985) Amphibia-Reptilia 6, 11-2.
Ragghianti, M., Bucci-Innocenti, S., and Mancino, G. (1978) Caryologia (Firenze) 31, 243-256.
Ragghianti, M., Bucci, S., Mancino, G., Lacroix, J.-C., Boucher, D., and Charlemagne, J. (1988) Chromosoma (Berl.) 97, 134-144.

Ragghianti, M., and Wake, D.B. (1986) Herpetologica 42, 206-214.
Rudak, E., and Callan, H.G. (1976) Chromosoma (Berl.) 56, 206-214.
Sanchíz, B., and Mlynarski, H. (1979) Acta Zool. Crac. 24, 175-188.
Schoorl, J., and Zuiderweijk, A. (1981) Amphibia-Reptilia 3/4, 235-252.
Schmid, M., Olert, J., and Klett, C. (1979) Chromosoma (Berl.) 71 29- .
Schmid, M., and Haaf, T. Origin and evolution of sex chromosomes in Amphibia: the cytogenetic data (preprint to G.M. 1988).
Sims, S.H., Macgregor, H.C., Pellat, P.S., and Horner, H.A. (1984) Chromosoma (Berl.) 89, 169-185.
Sites, J.W., jr., Chesser, R.K., and Baker, J. (1988) Copeia 1984(4), 1043-1053.
Sparreboom, M., and Arntzen, S. (1987) Proc. 4th Ord. Meet. S.E.H. Nijmegen (van Gelder, J.J., Strijbosch, H., and P.J.M. Bergers, Eds), pp.369-372.
Spurway, H. (1953) Symp. Soc. Exptl. Biol. 7, 200-237.
Thorn, R. (1969) Les Salamandres d'Europe, d'Asie et d'Afrique du Nord. Lechevalier, Paris, pp.1-376.
Tunner, (1974) Z. Zool. Syst. Evol.forschung 12, 309-314.
Twitty, V.C. (1955) J. Exptl. Zool. 129(1), 129-148.
Twitty, V.C. (1961) In: Vertebrate speciation (Blair, W.F., Ed), Univ. Texas Press, Austin, pp. 415-459.
Vallée, L. (1959) Impr. Declume, Lons-le-Saunier, pp.1-95.
Varley, J.M., Macgregor, H.C., Nardi, I., Andrews, C., and Erba, H.P. (1980) Chromosoma (Berl.) 80, 475-491.
Vrijenhoek, R.C., Angus, R.A., and Schultz, R.J. (1977) Evolution 31(4), 767-781.
Wake, B.B. (1966) Mem. So. California Acad. Sci. 4, 1-111.
Wake, D.B., and Özeti, N. (1969) Copeia 1969, 125-137.
Wallace, H. (1984) Caryologia (Firenze) 37, 3-8.
Wallis, G.P., and Arntzen, J.W. (1989) Evolution 43(1), 88-104.
White, M.J.D. (1973) Animal cytology and evolution. 3rd Ed., Cambridge University Press, Cambridge, pp. 1-961.
White, M.J.D. (1978) Modes of speciation. W.H. Freeman and Co., San Francisco, pp. 1-455.
Zaslavsky (1963, 1967) See: White, M.J.D. (1978).
Zuiderweijk, A. (1986) Proc. 3th Ord. Gen. Meet. S.E.H. Prague (Rocek, Z., Ed), 679-684.
Zuiderweijk, A., and Bouton, N. (1987) Proc. 4th Ord. Meet. S.E.H. Nijmegen (van Gelder, J.J., Strijbosch, H., and P.J.M. Bergers, Eds), pp.453-458.

Cytogenetics of Amphibians and Reptiles
Advances in Life Sciences

THE CYTOGENETIC STUDY OF EIGHT PELOBATID TOAD SPECIES BELONGING TO TWO GENERA OF SCUTIGER AND OREOLALAX IN CHINA (PELOBATIDAE, ANURA)

Li Shushen (1), Fei Liang and Ye Changyuan (2)

(1) Department of Biology, Yunnan University, Kunming, and

(2) Chengdu Institute of Biology, Academia Sinica, Chengdu, Popular Republic of China.

SUMMARY: The Giemsa karyotype, Ag-NORs and C-banding in eight Chinese pelobatid toad species (*Scutiger boulengeri*, *S.glandulatus*, *S.pingwuensis*, *S.tuberculatus*, *Oreolalax pingii*, *O.rugosus*, *O.liangbeiensis*, *O.chuanbeiensis*) have been studied here. Based on these cytogenetic informations all species examined can be divided into two groups: (1) genus *Scutiger* possessing the 2p secondary constriction and Ag-NOR and (2) the genus *Oreolalax* with the 6q secondary constriction and Ag-NOR. However both the groups have the same diploid number (2n=26) and karyotypic formula (6+7), as well as one pair of Ag-NOR commonly. Heteromorphisms of the secondary constiction, of the relative length of chromosome 1 and of the Ag-NOR have been observed. No dimorphic sex chromosomes have been discovered in the examined species. The reciprocal translocation and pericentric inversion probably existed in the chromosomal evolution of the two genera studied.

INTRODUCTION

The family of Pelobatidae may be a relatively ancient group in Anura and is generally considered as a transitional one between the primitive and advanced anurans. Pelobatids have an extensive fossil history extending from late Cretaceus of north America and Asia and from the middle Eocene of Europe through the Pleistocene. They occur in Europe, Asia and North America and have been divided into three subfamilies. The Megophryinae are distributed from East-Pakistan to Western China and through the Indo-Australian to the Sunda islands and Phlippine islands. The Pelobatinae live in Europe, Western Asia, North Africa and Southern North America to Southern Mexico. The Leptobranchinae occur in Western China, Nepal, North Burma and South East Asia (Dubois 1983, Duellman and Trueb 1985). The species belonging to Leptobranchinae are common in China especially in its South-western plateau and mountain where 50 species are recorded. The Trans-Himalaya mountain probably is their differentiation center. Research on their classification, ecology, habit, morphology have been made by some Chinese herpetologists early, but so far few taxonomic study have been provided. So this kind of research for this area is of importance to discuss the systematic and evolution.

For some genera such as *Leptobrachium*, *Megophrys*, *Leptolalax*, distributed in South-Eastern Asia, and *Pelobates*, *Scaphiopus* inhabiting in Southern North America and Mexico and Europe, the karyotypes and their evolution were reported and analysed by Morescalchi et al. (1977) and Wasserman (1970,

1973). Fresh data about their banding pattern were provided by Schmid et al. (1987). But as regards the Chinese pelobatid toads only very sparse available data have been recorded. The Giemsa karyotypes of <u>Oreolalax</u>, <u>Scutiger</u>, <u>Megophrys</u>, <u>Vibrissaphora</u> were reported by Zhao et al. (1983), Wu (1987) and Wu et al.(1988). So that we will provide the detailed karyotype and banding pattern of some pelobatid toad species in China in present paper.

MATERIAL AND METHODS

The studied animals were collected from the following localities as shown in table 1.

The mitotic chromosomes were prepared by means of method of bone marrow cells reported by Wu (1982). The animals were injected intraperitoneally or hypodermically with the Colchicine solution in dosage of 5-8 ug/g body weight, 6-12 h before killing. A bone marrow suspension was collected on a clean slide. The hypotonic treatment and the fixation were made in a moist chamber. The fixative was absolute ethanol, acetic acid and distilled water in the ratio of 1:2:3. The bone marrow cells were fixed for 2-2.5 h at room temperature. After air drying the slides were stained in 10% Giemsa in phosphate buffer pH 6.8 for 10-15 min. or were stored for banding treatment.

The technique of C-banding followed the method of Sumner (1972) with some modifications. The basic steps as follow:

(1) The slides, of which age did´nt exceed one week, were

rinsed in 0.2 N HCl solution at room temperature.

(2) Incubation in freshly filtered in 5% Ba(OH) for 20-30 min., according to the different species, at 50 C.

(3) Cleaned in 0.2 N HCl and 70% alcohol in sequence for discarding the adhesive sediment of Ba(OH), on the slide.

(4) Incubation in 2X SSC solution for 120 min. at 65 C.

(5) Rinse in distilled water followed by air drying, then stain in 10% Giemsa solution for 10-15 min.

TABLE 1. ANIMAL STUDIED

Species	Collection locality	specimens (m/f)
Scutiger glandulatus (Liu)	Zhongdian County Yunnan Province.	7/5
S.tuberculatus Liu et Fei	Puxiong County Sichuan Province	7/2
S.pingwuensis Liu et Tian	Pingwu County Sichuan Province	5/1
S.boulengeri (Bedriaga)	Pingwu and Nanping County, Sichuan Prov.	12/1
Oreolalax chuanbeiensis Tian	Pingwu County Sichuan Province	6/1
O.liangbeiensis Liu et Fei	Puxiong County Sichuan Province	9/2
O.pingii (Liu)	Caojiao County Sichuan Province	19/10
O.rugosus Liu	Caojiao County Sichuan Province	6/3

The Ag-NOR used the method of 1-step developer described by Howell and Black (1980)

The parameters for analysing karyotypes are the relative length, the arm ratio and the presence of secondary constriction. Based on the arm ratio the chromosomes can be distinguished into four types standardized by Levan (1964). The abbreviations of that are M=metacentric, SM=submetacentric, T=telocentric, ST=subtelocentric.

RESULTS

Giemsa stained karyotype: All species studied in the present paper have 2n=26 , of which two groups can be distinguished: 6 pairs of larger and 7 pairs of smaller chromosomes, i.e. the karyotypic formula is 6+7 (table 2). However the location of centromere and the relative length of the corresponding chromosomal pairs among species of a genus, as well as between the two genera are different. For example the no. 5 of *Scutiger* is ST, but in *Oreolalax* it is SM or M. No 5 in *O.liangbeiensis* is M but in the remaining species of *Oreolalax* it is SM. But as a whole the differences between the two genera are greater than those between the species in a genus. The eight species examined here can be divided into two groups, according to the different location of the secondary constriction which in *Scutiger* is located on 2p, whereas in *Oreolalax* is on 6q.

Heteromorphism in such character is also observed, that means presence of the secondary constriction on only one homologous chromosome. Dimorphism in the length of the long arm of chromosome 1 has been discovered in some species (Fig. 1). The long arm in one homolous chromosome is a little longer than its partner, but the additive segment is not heterochromatic after staining with the C-banding technique. Polymorphism in the diploid number has been observed in *O.liangbeiensis* were some metaphases have 2n=28. In comparison with the 2n=26 in this formula two additional small chromosomes are present.

TABLE 2: The karyotypes of the eight Chinese pelobatids

SPECIES	1	2	3	4	5	6	7	8	9	10	11	12	13
S.glandulatus	M	M	M	M	ST	M	M	M	M	M	M	M	M
S.tuberculatus	M	M	M	M	ST	M	M	M	M	M	M	M	M
S.pingwuensis	M	M	M	M	ST	M	M	M	M	M	M	M	M
S.boulengeri	M	M	M	M	ST	M	SM	M	M	M	M	M	M
O.chuanbeiensis	M	SM	SM	SM	SM	M	M	M	M	M	M	M	M
O.liangbeiensis	M	SM M	SM M	SM M	M	SM M	SM M	M	M	M	M	M	M
O.pingii	M	M	SM	SM	SM	SM	SM M	M	M	M	M	M	M
O.rugosus	M	M	M	SM M	SM	SM M	M	M	M	M	M	M	M

Ag-NOR: Only one pair of Ag-NORs appears in the eight species of the two genera. But there are two types of location. As to the four *Scutiger* species the Ag-NOR is located on the short arm of no. 2, whereas for the species of the genus *Oreolalax* it is on the long arm of no. 6. Therefore the localization of Ag-NOR just conforms to that of secondary constriction. As a general rule the occurrence of Ag-NOR is more stable in all metaphases observed (Fig. 2).

C-banding: for only five among the eight species studied we obtained satisfactory C-banded slides (Fig. 3). From the figure we can see that all the species possess a C-banding positive

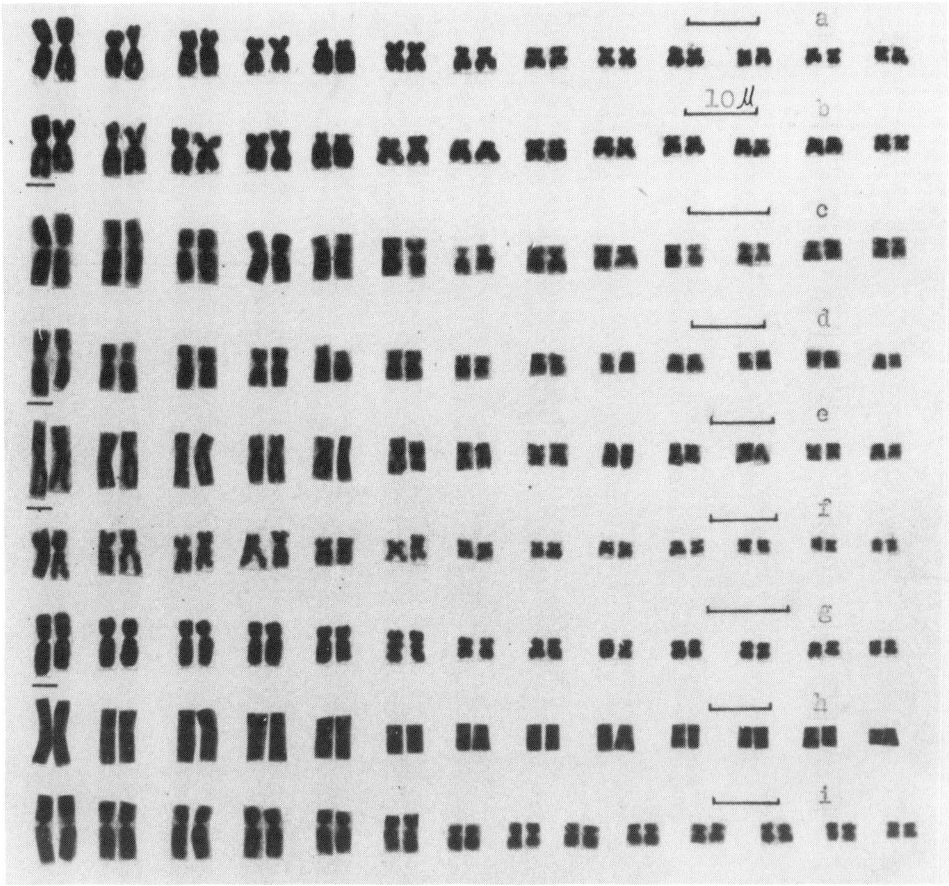

Fig.1 The conventional karyotype of eight Chinese pelobatids. a)*S.glandulatus*, b)*S.tuberculatus*, c)*S.boulengeri*, d)*S.pingwuensis*, e)*O.chuanbeiensis*, f)*O.pingii*, g)*O.rugosus*, h)*O.liangbeiensis*, (2n=26), i)*O.liangbeiensis* (2n=28). The line under the chromosome 1 indicates heteromorphism of long arm.

centromere , but in some of the smaller chromosomes of the genus *Oreolalax* the C-bands are very weak. No interstitial or telomeric C-bands can be seen. However a dense positive C-band occurs in the 6q of the three *Oreolalax* species where the secondary constriction is located. The C-banding positive block of O.liangbeiensis is rather large occupying almost 2/3 of the whole arm, the same is in O.pingii, while O.rugosus differs from the above two species showing weaker and smaller C-bands. A weak staining appears in the 2p where the secondary constriction of the genus Scutiger is located.

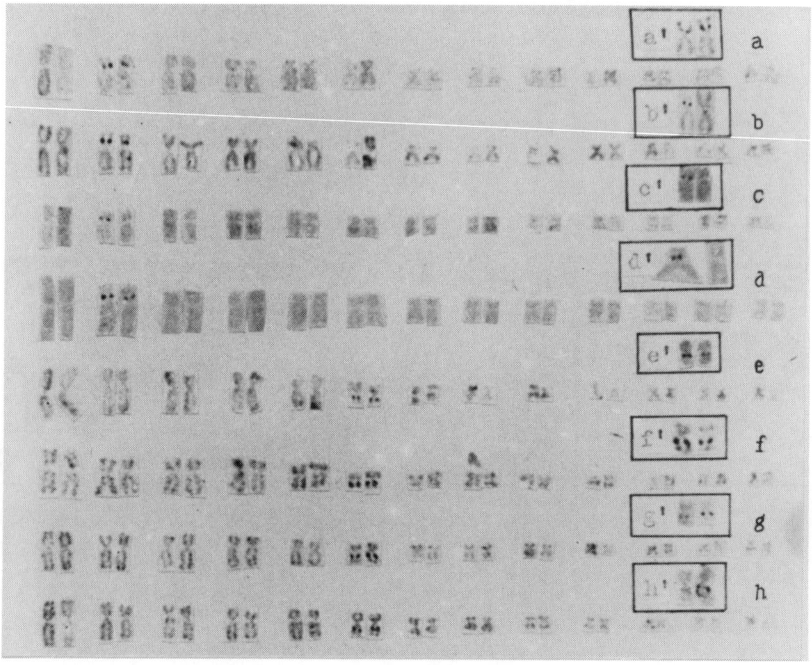

Fig.2 The Ag-NOR and its heteromorphism of the species studied. a,a´)S.glandulatus, b,b´) S.tuberculatus, c,c´)S.boulengeri, d,d´)S.pingwuensis, e,e´) O.chuanbeiensis, f,f´)O.pingii, g,g´) O.rugosus, h,h´)O.liangbeiensis.

As a whole the karyotype analysis and the differential banding techniques used here do not reveal the existence of sex-chromosomes.

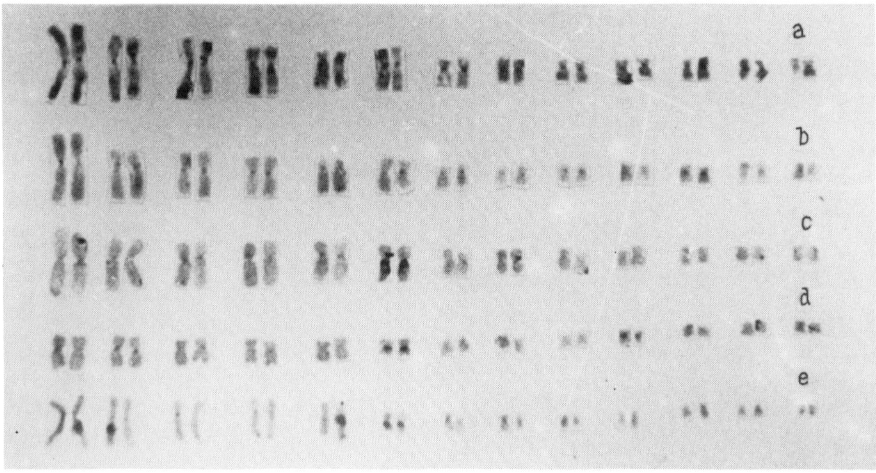

Fig.3 C-banding of five pelobatid species. a) *S.glandulatus*, b) *S.tuberculatus*, c) *O.liangebeiensis*, d) *O.pingii*, e) *O.rugosus*.

DISCUSSION

In the revision of Dubois (1983) the extant pelobatids include three subfamilies: Pelobatinae, Megophryinae and Leptobranchinae. According to cytogenetic studies most of the species of the three subfamilies have 2n=26, except some species of *Leptolalax* with 2n=24. As concerns the karyotypic differentiation within the pelobatids with 2n=26, two karyotypic formulae can be distinguished, the first one consisting of 6 pairs of larger and 7 pairs of smaller chromosomes (6+7) and the second consisting of 5 pairs of larger and 8 pairs of smaller chromosomes (5+8). The 6+7 formula is present in most species of Leptobranchinae as seen in the genera *Scutiger* and *Oreolalax* studied by us and *Vibrissaphora* studied by Zhao et al. (1983). Also in the Pelobatinae the species of the genus *Scaphiopus* (subgenus *Scaphiopus* only) and of the genus *Pelobates* possess this kind of formula. The 5+8 formula is present in most species of Megophryinae and also in species of *Scaphiopus* (subgenus *Spea*). The former formula has been considered as primitive as compared with the latter one (Morescalchi 1973, Morescalchi et al. 1977).

The species studied by us of both the genera *Scutiger* and *Oreolalax* have 2n=26 and the 6+7 foemula. All the chromosomes are biarmed including M, SM and ST, and the NF is 52. This differs from the pelobatids belonging to the subfamily Megophryinae where uniarmed chromosomes are present in various number, as for example in *Brachytarsophrys carnenis* (Tan et al. 1987), *Megophrys lateralis* (Wu et al. 1987), *Leptobrachium hasselti* (Morescalchi et al. 1977). As a general rule the biarmed chromosomes are considered as derived from uniarmed chromosomes, so the formula 5+8 may be relatively more primitive than the 6+7 formula. This is in contrast with the viewpoint of Morescalchi and coworkers (1973, 1977), but agrees with the paleontological information which demonstrates that the Megophrynae pelobatids are quite relative to ancient fossil species.

In the original classification before 1983, the extant pelobatids included only two subfamilies: Megophryinae and Pelobatinae. After that time the subfamily Megophryinae was divided into two subfamilies: Megophryinae and Leptobranchinae, of that the former may be a primitive group on the basis of fossil materials, morphological features and zoogeographical evidence. Within the subfamily Leptobranchinae the species examined by us especially those of the genus *Scutiger* are considered specialized.

Comparing the karyotypes of the two genera *Scutiger* and *Oreolalax* significant differences can be discovered. The former have one pair of ST (no.5) and a 2p secondary constriction and Ag-NOR, whereas the latter laks of ST chromosomes and has a 6q secondary constriction and Ag-NOR. On that condition these two taxa can not be considered as a single genus, as described by Duellman and Trueb (1985), but it is necessary to divide them into two independent genera. Recently this viewpoint has been supported also by morphological evidence (Fei and Ye, 1987).

Within the subfamily of Leptobranchinae, the 6q secondary constriction and Ag-NOR is frequent such as in *Oreolalax*, *Vibrissaphora* and *Leptobrachium*, and it is present also in some outgroup of Anura, e.g. *Amolops* species and some *Racophorus* species. Therefore this cytotaxonomic character may be plesiomorph, while the 2p secondary constriction and Ag-NOR occurring in all the Scutiger species, may be considered as apomorph. So that *Scutiger* may be more advanced in comparison to *Oreolalax*. This viewpoint conforms with the results of morphological research.

Within the genera *Scutiger* and *Oreolalax*, although the 2n and the secondary constriction and Ag-NOR position are similar in all the species of the same genus, the relative length and arm ratio in some corresponding chromosome pairs display significant differences. Therefore reciprocal translocation and pericentric inversion may be infered as the main karyological evolutionary mechanisms in these frogs.

A particular aspect is the heteromorphism in the two homologous of the chromosome 1, where the long arm of one homologous is generally little longer than that of the other. This kind of heteromorphism has been reported in crested and marbled newts (*Triturus*) (Sims et al. 1984) on which difference not only appears in the length of the two homologous, but also the additional segment is C-banding positive. This signifies that the additional segment is made up of constitutive heterochromatin. On the contrary such segments are C-banding negative in the species studied by us.

The C-banding reveals that in all species examined the C-banding material is less developed compared with other anurans like the genera *Rana*, *Amolops*, and so on. They are without visible interstitial and telomeric C-bands, except four species of the genus *Oreolalax*, in which the more dense C-banding block has an interstitial location. So the condition of C-banding for these species display a primitive characteristic. The location of the interstitial C-banding in *Oreolalax* just conforms with that of secondary constriction and Ag-NOR. On the contrary the location of seconadry constriction and Ag-NOR on 2p in *Scutiger* species is C-banding negative or shows only a very weak staining. Such phenomenon was reported in early articles on Amphibian cytogenetics and showed the existence of a heterogeneity in the Ag-NOR or secondary constriction. The study of Matsui et al. (1986) using the N-banding method to distinguish the subregion of the NOR, revealed the existence of two distinct non-histone proteins relative to the Ag-NOR.

Just like most of the pelobatid toads, except in Leptobrachium hasselti (Morescalchi et al. 1977), there is only one pair of Ag-NOR in the eight species we studied. The location of Ag-NOR is different in the two genera of Scutiger and Oreolalax. The former is on 2p, whereas the latter is on 6q. No visible difference appears between the species within the same genus. So the localization of the Ag-NOR displaies a generic specificity. The characteristices of Ag-NOR are more stable than those of the secondary constriction. The latter always presents variations due to experimental conditions. The presence of the Ag-NOR generally is constant in each metaphase although they have dimorphism phenomena. So it is a reliable character in discussing the problems of cytotaxonomy. Accumulating information shows that the generic level of Pelobatidae has special Ag-NOR properties. The genus Scutiger has the Ag-NOR on 2p, Oreolalax on 6q, Vibrissaphora on 6q also but without positive C-banding (Li et al., to be published), Pelobates on 7p, Megophrys on 6q, Scaphiopus on 1p, in the subgenus Scaphiopus, and on 6q, in the subgenus Spea (Schmid et al. 1987). From this point the genus Scaphiopus may be divided into two genera. It is quite obvious that there are no visibe differentiation of Ag-NOR within the genera we studied.

Ag-NOR heteromorphism has been observed in most species we studied here. There are two situations in Ag-NOR heteromorphism, the former is the size difference of

Ag-labelling on the two homologues, the Ag-block on one homologous being larger than that on the other one; the second situation is the presence or absence of Ag-labelling on the homologous chromosome pair. Such intraindividual heteromorphism of Ag-NOR of Amphibia has been frequently described in various taxa of both primitive and advanced anurans (Ward 1977, Schmid et al 1987). According to the viewpoint of Schmid et al. (1987) this phenomenon is induced by the tandem duplication or deletion of 18S+28S RNA genes due to the unequal meiotic crossover. It indicates that the number of the 18S+28S RNA coding genes is different between the homologous partner chromosomes.

CONCLUSION

Although the 2n (26), the karyotypic formula (6+7) and the number of Ag-NORs are the same between the two genera *Scutiger* and *Oreolalax*, however its location of secondary constricion and Ag-NOR is different. The former is on 2p, whereas the latter on 6q. Considering that they should be rank as two independent genera. Besides there are significant difference among species in the relative length and arm ratio of some corresponding chromosomal pairs wether within the genus or between genera. Therefore the reciprocal translocation and pericentric inversion are probably their main karyological evolutionary mechanisms. Heteromorphism in Ag-NOR and in the length of chromosome 1 has been discovered. The additional segment of long arm displaies a negative C-banding reaction.

Aknowledgments: This study is a part of subsidized item from the fund committee of National Nature Science in China. We are grateful to Luo Jiarui and Zhu Baosen who are graduate students of Biology Department in Yunnan University for collecting animals and some laboratory works.

REFERENCES

Dubois, A. (1983) Bull.Mem.Soc. Linnenne Lyon 52, 270-276.

Duellman, W.E., and Trueb, L. (1985). Biology of Amphibians, Mc Graw-Hill, New York, st. Louis, San Francisco, pp. 522-524.

Fei Liang and Ye Changyuan (1987) Acta Biologica Plateau Sinica 7,155-170 (In Chinese with English summary)

Howell, W. and Black D.A. (1980) Experientia 36, 1014-1015

Matsui,S., Fuke,M., Elassoul, S., Chai,L., and Sandberg A.A. (1986) Chromosoma 93,231-242.

Morescalchi, A. (1973). In: Cytotaxonomy and Vertebrate evolution (Chiarelli, A.B. and Capanna E. Eds.)Academic Press, London, New York, pp.270-273.

Morescalchi,A., Olmo, E., and stingo, V. (1977). Experientia 12, 1577-1578.

Schmid, M., Vitelli L., and Batistoni R. (1987) Chromosoma 95, 271-284.

Sims, S.H., Macgregor, H.C., Pellatt, P.S., and Horner H.A. (1984). Chromosoma 89, 169-185.

Tan Anming, Zen Xiamao, Wu Gaunfu, and Zhao Ermi (1987). Acta Herpetologica Sinica 1, 1-4 (In Chinese with English summary).

Ward O.G. (1977) Can.J.Genet.Cytol. 19,51-57.

Wu Guanfu, Tan Anming, and Zhao Ermi (1988) Acta Herpetologica Sinica (In Chinese, with English summary).

Wu Guanfu (1987) Acta Herpetologica Sinica 3, 45-48 (In Chinese with English summary).

Wu Zhengan (1982). Hereditas (Beijing) 1, 38-39 (In Chinese).

Zhao Ermi, Wu Guanfu and Yang Wenning (1983) Acta Herpetologica Sinica 1, 20-23 (In Chinese with English summary).

Cytogenetics of Amphibians and Reptiles
Advances in Life Sciences

FLOW CYTOMETRIC DETERMINATION OF CHANGES IN CELLULAR DNA CONTENT DURING DEVELOPMENT OF *XENOPUS LAEVIS*

B. Fritz*, J. P. Fritz+ and W. Ulrich*

*Institut für Genetik , Universität Hohenheim, D-7000 Stuttgart 70, Federal Republic of Germany +Zoologisches Forschungsinstitut und Museum Alexander Koenig, Adenauerallee 150-164, D-5300 Bonn 1, Federal Republic of Germany

SUMMERY: Changes in relative cellular DNA content in tadpoles of *Xenopus lavis* of different developmental stages had been observed by flow cytometry. The DNA amount increased from stage 46 to 50 and decreased after stage 57 to the normal level of adult animals. After overdosage of hormone for induction of spawning in adult *Xenopus* different mutations such as triploid, tetraploid, hexaploid forms and genetic mosaics had been detected in hatching tadpoles.

DNA analysis of monodispersed animal cells demonstrate characteristic DNA distribution curves with peaks for nonreplicating cells (cells in the G_1/G_0 phase of the cell cycle), usually expressed as 2C cells and postreplicated and mitotic cells (G_2+M or 4C cells). Between both peaks the amount of cells in the S phase (DNA replication) is recognizable. Alterations of cellular DNA content can be detected with great exactness by the FCM technique. Changes in the nucleic acid concentrations during embryonic development of Xenopus laevis have been reported (Bristow & Deuchar, 1964). These findings derived from biochemical spectrophotometric data. In this investigation flow cytometric DNA analysis was done of Xenopus laevis tadpoles of different developmental stages after hormone induced spawning.

MATERIALS AND METHODS

Animals: Xenopus laevis tadpoles of different developmental stages (Nieuwkoop & Faber, 1956). These animals have been obtained by hormone application (choriongonadotropine; 250 units/animal) to induce spawning in parental animals .

Cell preparation and staining: described by Ulrich et al., 1988 previosly.

Flow cytometric measurement and analysis: Measurements were done with a PAS-II impulscytophotometer (Fa. Partec AG) using combinations of UG1, TK 420, TK560 and GG435 (for DAPI) and RG590 filters (for SR101). The DNA content of analyzed tadpole cells was calculated on the basis of their ratio to simultaneous stained chicken red blood cells (CRBC).

RESULTS AND DISCUSSION

Flow cytometry is a simple, rapid and sensitive cytogenetic method for cellular DNA analysis in animal cells. The use of the fluorochrome combination DAPI/SR101 yield accurate DNA distribution curves (DNA histograms) and alterations in cellular DNA content can be estimated up to 2%.

Flow cytometic DNA analysis of cells prepared from tadpoles of different embryonic developmental stages [35-58] showed an increasing of cellular DNA content starting at stage 46, reaching a maximum at stage 50 to stage 54 and than decreased after stage 57 to amounts of mature animals (Fig. 1).

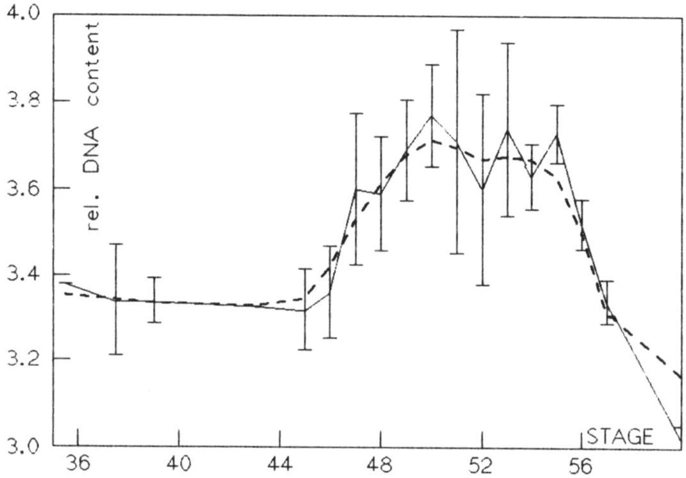

Figure 1 : Nuclear DNA amount of *Xenopus laevis* tadpoles at different stages of development. Abscissa : developmental stages (35-56) evaluated according to Nieuwkoop & Faber (1956). Ordinate : relative DNA amount expressed by the ratio of $G_{0/1}$ cells of the samples to the mode of $G_{0/1}$ reference cells (CRBC). The dashed line indicates a smoothed curve for nuclear DNA amount. Number of analyzed individuals n = 107.

Comparison of flow cytometric data and morphological characteristics of tadpole development conclude that loss of yolk salk and beginning of ingestion (at stage 45-46) correlates with increasing of cellular DNA. When organs are differentiated and extremities are developed (stage 57) the DNA content decreased on normal values. For interpretation of this phenomenon we suggest that an increasing of cellular DNA content is induced by gene amplification for specific gene expression to promote differntation during development. On the other hand decondensation of compact chromatin might be responsible for the higher DNA content data. There are some arguments against the use of DAPI for cytofluorometric analysis because of its adenine-thymine rich regions binding capacity

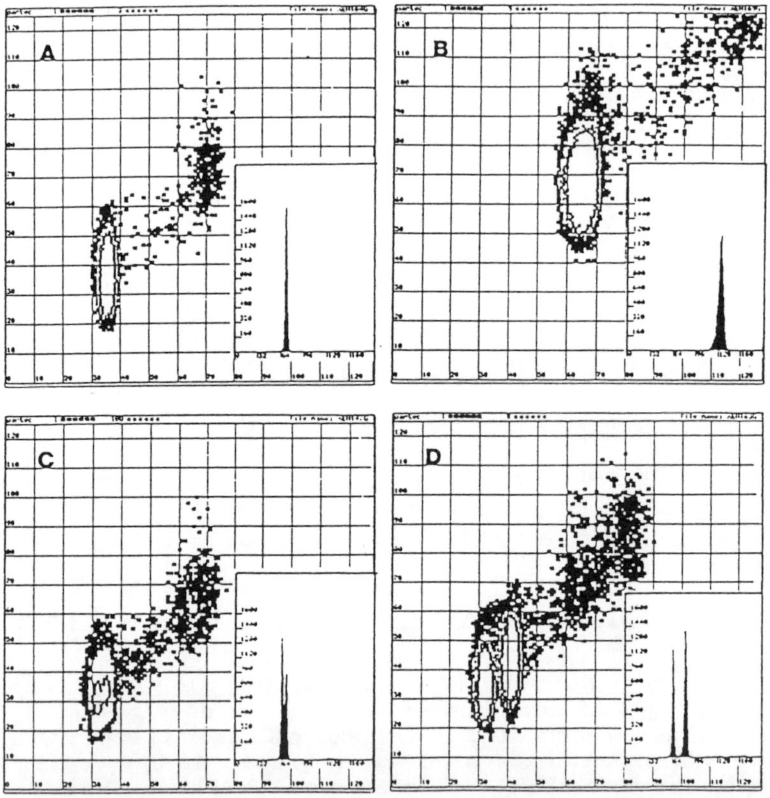

Figure 2: FCM-analysis of Xenopus laevis tadpoles obtained after hormone overdosage. Bivariate contour displays showing the correlation of DNA and protein content. Sequential contours at increasing isometric levels represent the distribution DNA vs. protein. Abscissa: relative DNA content. Ordinate: relative protein content. Univariate DNA distributions are inserted. Abscissa: relative fluorescence intensity (relative DNA content). Ordinate: cells/channel (relative cell number). A: histograms of normal diploid Xenopus laevis tadpole B-D: examples of abnormal DNA distibutions in Xenopus laevis cells. B: tetraploid modal DNA value, C + D: analysis detects subpopulations with different DNA content. Histograms showing at low grade DNA aneuploidy with DNA indices (Hiddemann et al., 1984). of 1.1 (C) and 1.3 (D).

to DNA and therefore its minor affinity to heterochromatic chromatin (Capriglione et al., 1987). Our experience in flow cytometric DNA analysis of animal and human cells point to an increasing of cellular DNA content in tadpole cells during embryonic development.

Table I: Frequency of genetic alterations in phenotypically altered <u>Xenopus laevis</u> tadpoles based on FCM DNA analysis

number of examinated tadpoles		diploid DNA content		total		GENOTYPIC ALTERATIONS									
						triploid		tetraploid		hexaploid		diploid + aneuploid		triploid + aneuploid	
n	%	n	%	n	%	n	%	n	%	n	%	n	%	n	%
46	100	25	54.3	21	45.7	6	13.1	3	6.5	7	15.2	3	6.5	2	4.4
												DNA INDICES			
												1.1-1.3		1.08-1.1	

These findings are in agreement with reported investigations of biochemically analyzed data of increasing nucleic acid concentrations in <u>Xenopus laevis</u> embryos (Bristow & Deuchar, 1964).

Inadvertently application of overdosage of the hormone chorion-gonadotropine for induction of spawning in parental animals produced phenotypic (e.g. incoordinated mobility, S-sharped curvature of the tail, reduced pigmentation)and genotypic changes in progeny (Table I). Flow cytometric DNA analysis of some viable tadpoles demonstrate alterations in cellular DNA (Fig. 2). Euploid DNA contents (triploid, tetraploid and hexaploid) and combinations of diploid or euploid cells with aneuploid cell populations were examined in single tadpoles.

<u>REFERENCES</u>

Bristow, D.A., and Deuchar, E.M. (1964) Exp. Cell Res. <u>35</u>, 580-589.

Capriglione, T., Olmo , E., Odierna, G., Improta, B., and Morescalchi, A. (1987) Bas. Appl. Histochem. <u>31</u>, 119-126.

Hiddemann, W, Schumann, J, Andreeff, M, Barlogie, B, Herman, C.J, Leif, R.C, Mayall, B.H, Murphy, R.F, and Sandberg, A.A (1984) Cytometry <u>5</u>, 445-446.

Nieuwkoop, P.D., and Faber, J. (1956) N. Holland Publ. Co. Amsterdam

Ulrich, W., Fritz, B., and Fritz, J. P. (1988) Bonn. zool. Beitr. <u>39</u>, 49-58

Cytogenetics of Amphibians and Reptiles
Advances in Life Sciences

KARYOTYPIC CHARACTERIZATION OF BROWN FROGS FROM THE IBERIAN PENINSULA (RANIDAE:RANA).

P. Herrero, B. Arano and M.L. Esteban

Museo Nacional de Ciencias Naturales de Madrid. C/ J. Gutiérrez Abascal 2, 28006 Madrid. Spain

SUMMARY: Three species of "brown frogs" (Rana temporaria, Rana iberica and Rana dalmatina) and one "green frog" species (Rana perezi) from the Iberian Peninsula have been cytogenetically analyzed using different chromosome banding techniques. Paleontological evidence points to R. perezi as closest to the ancestral morphotype of Rana found in the Peninsula. Based on this and on the chromosomal data R.iberica would clearly represent the most derived group among the brown frog species studied, with R. temporaria being its closest related species. The different distribution of constitutive heterochromatine between iberian and central european populations of R. temporaria indicates a reproductive isolation corresponding to subspecific level.

Studies dealing with European brown frogs have been scarce. Most of them used only one or two species, usually belonging to the Rana temporaria complex, in comparing them with Northamerican "brown" and "stream" frogs (Green, 1986a, 1986b; Farris et al., 1979, 1982). Just recently, attempts have been made to define phylogenetic relationships among mediterranean brown frogs based on electrophoretical and morphological evidence (Giacoma, pers. com.). However, no comparative cytogenetic studies have been carried out on this group.

The purpose of this paper is to present with a comparative cytogenetic analysis of the three species of brown frogs present in the Iberian Peninsula based on identification of chromosomes, as well as characterization of certain chromosomal regions by means of differential staining of both constitutive heterochromatic and nucleolus organizer regions (NORs). The study comprises the species R. iberica, R. temporaria and R. dalmatina, including R. perezi as representative of green frogs.

MATERIALS AND METHODS

The localities, sex and number of the specimens are listed as follows:

Rana temporaria ----------------------	Lago Isoba (2 ♀, 8 ♂), Gijón (2 ♀, 2 ♂), Villaviciosa (4 ♀, 4 ♂), Lagos de Saboredo (10 ♀, 10 ♂), El Portalet (6 ♀, 6 ♂)
Rana iberica -------------------------	Laguna Grande de Gredos (10 ♀, 10 ♂) Segovia (20 juv.), Mijares (3 ♀, 4♂) La Estrada (2 ♀, 3 ♂), Moal (2 ♀, 3 ♂)
Rana dalmatina -----------------------	Irati (6 ♀, 7 ♂)
Rana perezi --------------------------	Salamanca (3 ♀, 7 ♂), Dehesa de Campoamor (9 ♀, 8 ♂)

All the frogs used in this study are deposited in the National Museum of Natural History in Madrid.

Chromosomes were obtained using gut epithelium and testes as described in Herrero (1982).

C-banding, DAPI-DA and CMA -DA differential staining procedures were carried out according to Herrero et al. (1989).

The idiograms were constructed following the description in Herrero (1982).

RESULTS AND DISCUSSION

All four Iberian species have 2n=26 chromosomes, which is the typical diploid number of most Rana species. The chromosomes of each karyotype can be divided into two size groups: five pairs of large chromosomes and eight pairs of small chromosomes (Fig. 1a,e,i). Prominent secondary constrictions associated with NORs, as revealed by silver staining (Fig. 1b,f,j), occur in chromosome 10 of R. temporaria, R. iberica and R. perezi. On the contrary, secondary constrictions occur on the short arms of chromosomes 3 and 10 of R. dalmatina as reported by Guillemin (1967). The results obtained for this species, though not entirely clear, point to the fact that the secondary

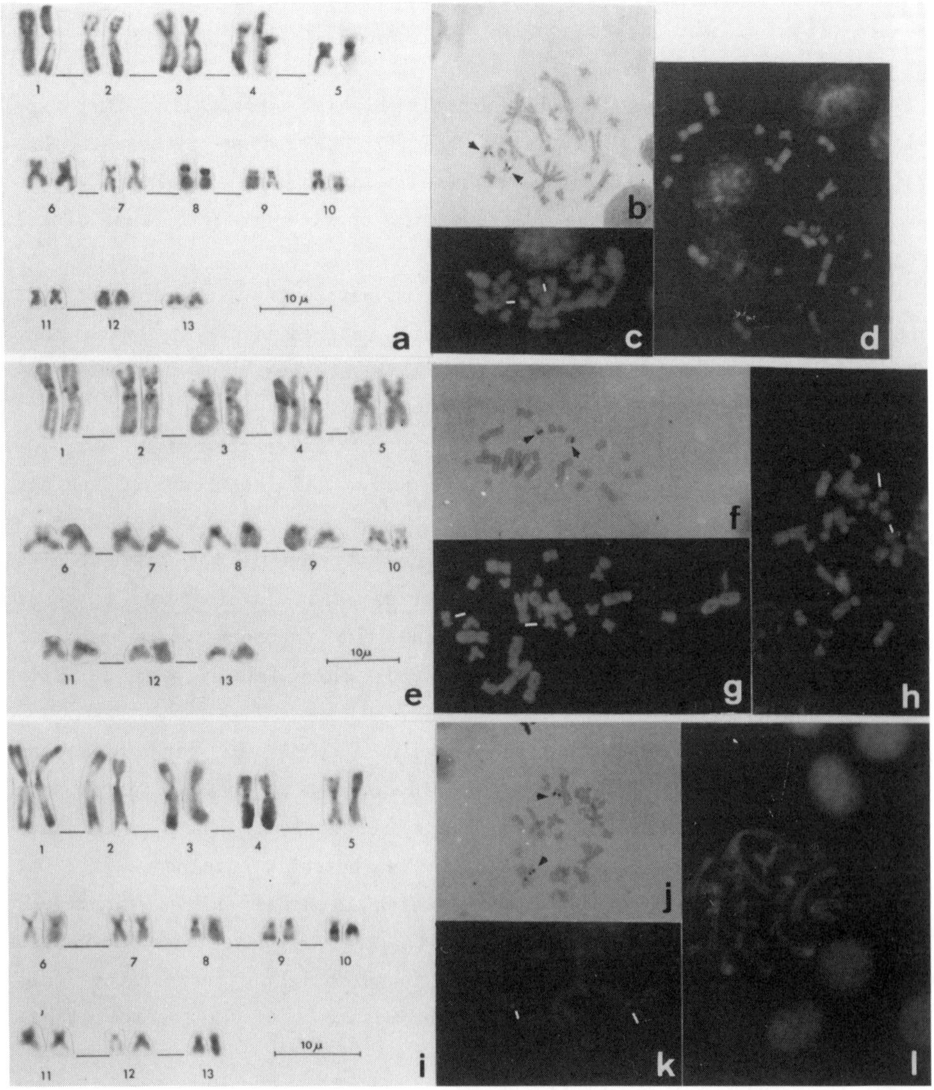

Figure 1: a,b,c,d Chromosomes of Rana temporaria. e,f,g,h. Chromosomes of Rana iberica. i,j,k,l Chromosomes of Rana perezi. a,e,i C-banded karyotypes. b,f,j Silver stained metaphases and NOR locations. c,g,k CMA /DA banding showing a positive band in chromosomes 10. d,h,l DAPI/DA banding. (h) Metaphase showing a positive band in chromosomes 8 of R. iberica.

constriction of the short arm of chromosome 10 could be associated to a NOR.

Although the species of *Rana* exhibit highly conservative karyotype morphology, many of them show specific C-banding patterns. That is the case of *R. perezi* which has few and faintly stained C-bands exclusively located in the centromeric, pericentromeric and distal regions (Figs. 1i,2).

R. temporaria has almost all C-bands in pericentromeric, interstitial and distal positions (Figs. 1a,2). This pattern contrasts with that shown by German individuals of the same species. In this case C-bands only appear in the centromeres (Schmid, 1978).

R. iberica shows the highest amount of heterochromatin with C-bands concentrated in the pericentromeric and centromeric regions (Figs. 1e,2).

The similarities between the C-banding patterns of the four species analyzed is very restricted: *R. iberica* and *R. temporaria* share the same pattern in chromosome 8, and the presence of a band in chromosome 10 where the NOR is located. This latter band is also exhibited by *R. perezi* (Fig. 2).

The use of Chromomycin A_3 /Distamycin A (CMA_3/DA) and Diamidine Phenylindole/Distamycin A (DAPI/DA) in order to reveal chromosome regions rich in GC and AT base pairs, respectively, has shown that the C-band associated with the NOR is brightly stained with CMA_3 /DA in all species (Figs. 1c,g,k,2), which is usual in the karyotypes of lower vertebrates. No other GC-rich regions have been revealed. Regarding DAPI/DA banding, we have found bright fluorescence only in a thin band in the short arm of chromosome 8 of *R. iberica* (Figs. 1d,h,l,2).

The results suggest that all three species of brown frogs are readily discriminated and distinctive in their chromosome characteristics. Some predictions about their phylogenetic relationships can also be made, especially concerning the two species *R. dalmatina* and *R. iberica*. According to the presence of an unusual secondary constriction in chromosome 10 of *R. dalmatina*, this species could be the most distant representative. As for *R. iberica* it would be representing the most derived state of the iberian species studied. Based on

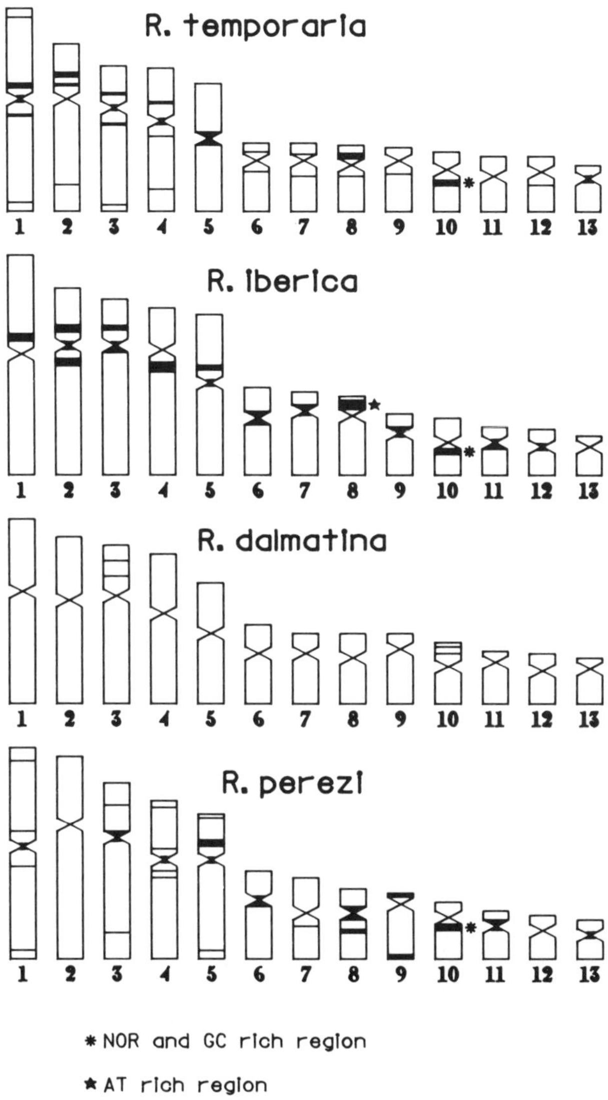

Figure 2: Idiograms of C-banded karyotypes of *Rana* species studied.

paleontological evidence, the earliest fossil remains of the genus Rana (closest related to the present group of green frogs) found in Europe date from the Lower Miocene, whereas the oldest brown frog-like fossils come from Pliocene deposits (Sanchíz, 1977). Therefore R. perezi can be considered the specied closest to the ancestral colonizers of the Iberian Peninsula.

The cytogenetic results show that this species has the lowest amount of heterochromatin of those groups examined, while R. iberica not only shows the highest amount of heterochromatin but is also differentiated by the presence of an AT base pair rich band which is exclusive of this species. All this strongly supports the hypothesis of R. iberica being the most recently derived of the groups studied.

Finally, the differences shown between German and Iberian populations of R. temporaria suggest that the geographic isolation of Iberian populations has allowed for an important divergence which is clearly revealed by cytogenetic analyses. The significance and extent of such a divergence should be tested by means of experiments on artificially bred hybrids.

These results presented stress the need for much broader studies on all Paleartic "brown frogs", especially on the so called R. temporaria complex where most of the variability is still unknown.

Acknowledgements: We would like to thank M. Vences and A. Gosá for providing many of the animals used in this study. This work was supported by CICYT PB880010 Project.

REFERENCES

Farris, J.S., Kluge, A., and Mickevich, M.F. (1979) Syst. Zool., 28, 627-634.
Farris, J.S., Kluge, A., and Mickevich, M.F. (1982) Syst. Zool., 31, 479-191.
Green, D.M. (1986)a Syst. Zool., 35, 273-282.
Green, D.M. (1986)b Syst. Zool., 35, 283-296.
Guillemin, C. (1967) Chromosoma, 21, 189-197.
Herrero, P. (1982) Herpetologica, 38, 502-506.
Herrero, P., Bella J.L., and Arano, B. (1989) Heredity 63,119-123.
Sanchíz, F.B. (1977) Univ. Complutense de Madrid. Tesis Doctoral. III vols., pp. 863.
Schmid, M. (1978) Chromosoma, 68, 131-148.

Cytogenetics of Amphibians and Reptiles
Advances in Life Sciences

KARIOLOGICAL INVESTIGATION OF THE BROWN FROGS SPECIES (ANURA - RANIDAE) FROM BULGARIA AND GREECE

R.G. Belcheva and Th.S. Sofianidou

Department of Genetics, Faculty of Biology, University of Sofia and Department of Zoology, University of Thessaloniki

SUMMARY: Three species of brown frogs, Rana dalmatina, R. graeca and R. temporaria from Bulgaria and Greece have been investigated karyologically. The chromosome set of the three species consists of 26 chromosomes, 10 large and 16 small. A sonsiderable variation in the structure of the secondary constriction was found. The specimens from all localities have constitutive heterochromatin at the centromer region which in some cases was extended to the telomers. B-chromosomes were found in the specimens of R. temporaria from one locality in the Balkan Range.

The large Ralearctic "brown frogs" group of Rana temporaria and its many allied species is distributed across Eurasia from Western Europe to Japan. Three species of this group, R. dalmatina, R. graeca, and R. temporaria, are components of Bulgarian and Greek founa. The karyotypic analysis by means of chromosonic banding technique has only been described in R. temporaria (Schmid, 1978). The present report summarizes some of the results of the karyological investications on the above three species from different localities of Bulgaria and North Greece. The analyses were performed with conventional cytogenetic methods as well as with differential staining of chromosomes.

MATERIAL AND METHODS

A total of 40 specimens of both seces of R. temporaria, R. graeca and R. dalatina, collected from several localities of Bulgaria and North Greece were karyotypecally analyzed.

C-and Ag-banding was done according to Summer (1972) and Goodpasture & Bloom (1975) respectively.

RESULTS AND DISCUSSION

The three specimens analyzed, like most members of the genus Rana have 26 chromosomes which are divided into a group of 5 large pairs and a group of 8 smaller pairs (figures 1.I, 2.I, 3.I). In one specimen of R. graeca chromosome pair 7 was heteromorphic and consisted of one submetacentric and one subtelocentric chromosomes of equal length. This can only be explained as a result of a pericentric inversion. If there is an advantage to such a polymorphism it must be due to the presence of isolated chromosomal segments for free crossing-over between homologues different centromeric location to apt to produce genetically imbalaced gametes. The 10th chromosome of R. temporaria and R. graeca has a secondary constriction in the long arm. Besides the secondary sonstriction in the 10th pair R. graeca has an additional secondary constriction on the short arm of chromosome 6 (fig. 4). In the karyotype of R. dalmatina the large secondary constrictions das not lie in the long arm of chromosome 10, but farther in the short arm of chromosome 3 (fig. 3). In this case it is possible that a reciprocal translocation took place between the chort arm of chromosome 3 and the long arm of chromosome 10 (Schmidt, 1976). There is no visible evidence of sex chromosome dimorphism in these three species.

NORs

Studies which have investigated the location of the NOR in Anurans one feature has consistently emerged: in closely related species the NOR is almost always localized in the same region of the same chromosome pair. Systematic studies on a large number of species have showh that most of the primitive and higly evolved Anura present only on pair of Ag-staind NORs in their Karyotypes (Scmid, et al., 1986). It has been suggested that karyotypes with on sigle homologous pair of NORs are more primitive than those in which the NORs are scattered over several chromosomes (Hsu et al., 1975).

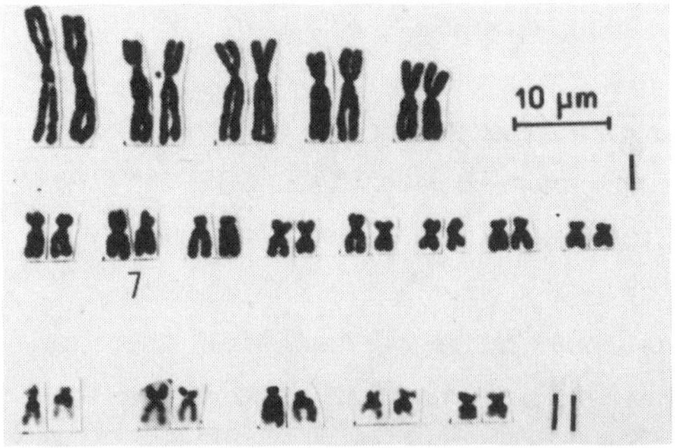

Fig.1

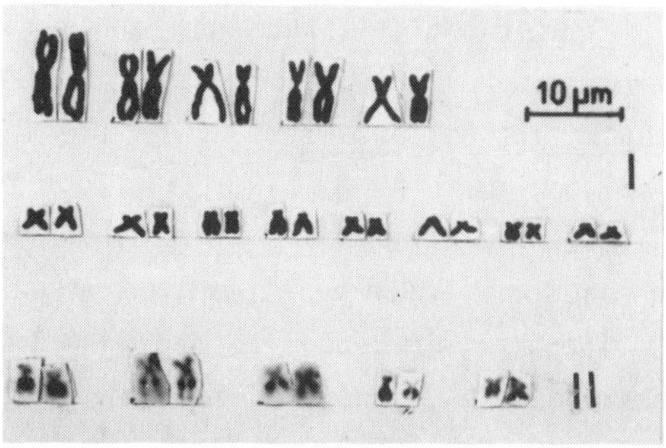

Fig.2

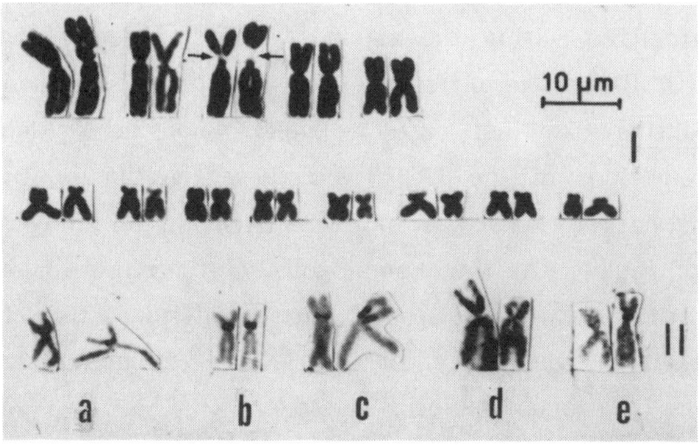

Fig.3

Fig.1. I-Karyotype of Rana temporaria. II-Selected NOR-carrying chromosome pair 10th from five specimens after Ag-staining. Note the different sizes of Ag-blocks.

Fig.2. I-Karyotype of Rana graeca. Note that pair 7th consistes of one submetacentric (left) and one subtelocentric (right) chromosome of equal length. II-Selected chromosome pairs from one metaphase showing different site of Ag- staining.

Fig.3. I-Karyotype of Rana dalmatina. Note the unequal sizes of the secontary constrictions (arrows) in chromosome pair 3th. II-Selected NOR-carrying chromosome pair 3th from five specimens after Ag-staining. Note the remarkably great interindividual variabiliy in the lengths of the Ag--stained blocks. a,d, exmble of unequal-sized homologous NORs. b, example of homologous of equal sizid NORs. c, two silver stained chromosome pair shoving a duplicated nucleolus organizer in one chromosome (right) and a normal-structured nucleolus organizer in the homologous chromosome (left). e, a deletion of the NOR from one of the homologous chromosome (left) and a normal-structured nucleolus organizer in the homologous chromosome (right).

Figures 1-II, 2-II, 3-II show chromosome pairs possesing NORs in the three investigated Rana species after Ag-staining of the chromosomes. In R. temporaria and R. graeca NORs are located in the secondary constriction of the long arms of the 10th (10q) chromosome pair. Besides the NORs in the 10th chromosome, R. graeca has a few addetional sites in different chromosomes which were positively stained by Ag.

The size differences between the NORs could be recognized after specific staining with Ag. In one male of R. dalmatina with unequal sized NORs the different size was due to a tandem duplication in one of the two NORs. Two Ag-blocks were localized within the secondary constriction of both chromatids in this case (fig. 3, II-c). Size difference between homologous NORs, like those found in R. dalmatina examined, are frequend among the Amphibia (Schmid, 1982). Complete deletions of the NORs are rarer than duplication, but in one specimen of R. dalmatina there was only one NOR in all metaphases analysed. The corresponding region on the homologous chromosome naner exhibited a nucleiolar constriction nor did it react with Ag (fig. 3,IIe). It is generally assumed that the high variability of the NORs including duplications and deletions, originates by unequal meiotic crossover.

C-Band

The three Rana species have C-band heterochromatin hevily concentrated in centromeric positions and few interstitial C-bands. In one male of R. graeca heterochromatic blocks of two different sizes were seen in chromosomes 1, 2, 3 (fig 4a) in all cells examlined. The observed heteromorphism of heterochpomatin between homoloque chromosomes may be reflects a polymorphism within the popylation as each would have contributed different amounts of heterochromatin per homoloque. In one specimen in a few cells there were heterochromatic blocks only on one of the chromatids of four chromosomes (fig. 4c). This reflected unequal sister chromatid exceges in this region, and created possibilities of C-band polyporphism.

B-chromosomes

Among Anura only four spesies with B-chromosomes have been desribed up to date (for references see Jones and Rees 1982). Earlier Ullerich (1967) and Schmid (1978) reported supernumerary chromosomes in specimens of R. temporaria obtained from southern Germany and Central Europe. We found supernumerary B- chromosomes in all the nine specimens of R. temporaria analyzed from one

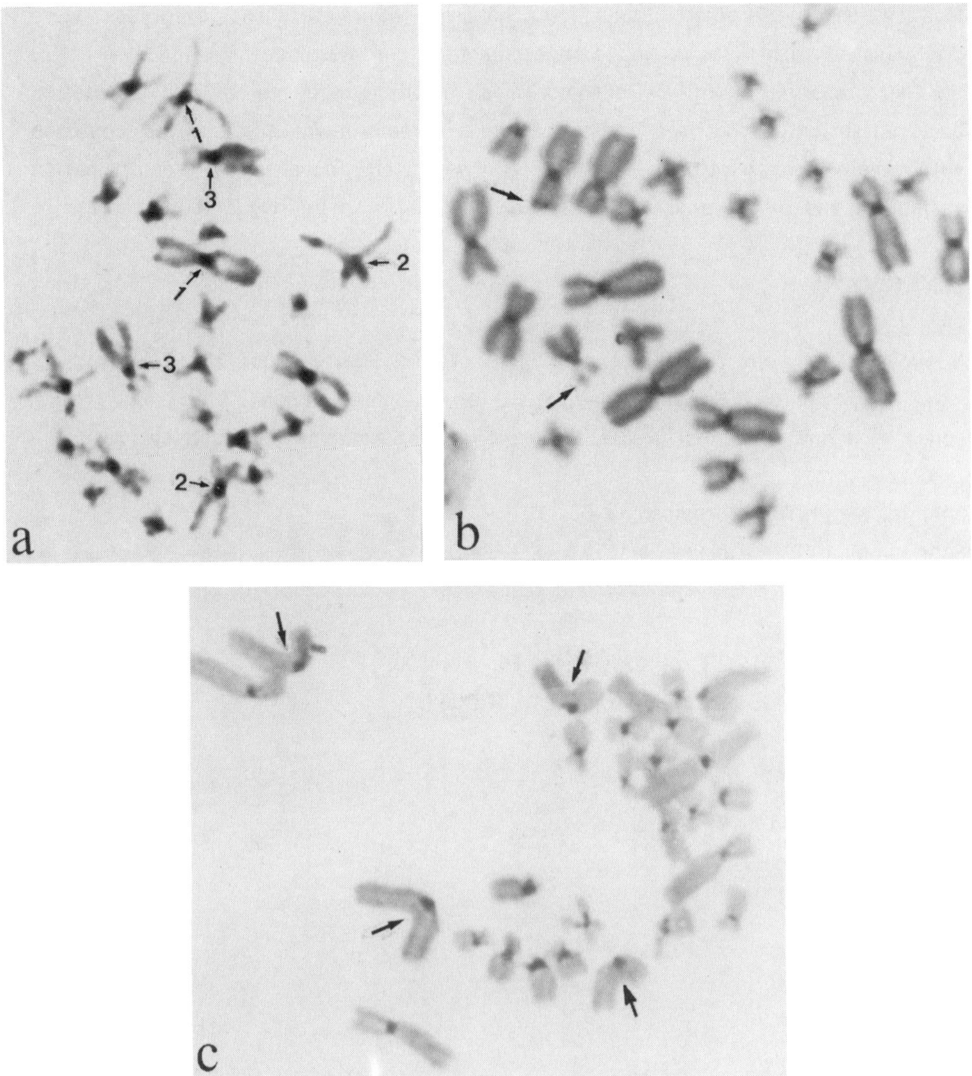

Fig. 4. C-baded metaphases in three specimens of Rana graeca. a, size difference between homologous heterochromatic bands in chromosomes pairs 1, 2, 3 (arrows) recognizable in all mitoses analyzed. b, a weak C-band in the telomere of the short arm of chromosome 4th (arrow). Note and a constriction in the short arm of chromosome 6. c, C-blocks only in one chromatids (arrows).

locality on the Bolkan Range in Bulgaria. Four different cytotypes were distinguished with 1, 2, 3, and 4 supernumerary chromosomes·

The preservation of B- chromosomes in the specimens of the population analysed inticates that they ensure them a certain advantage may be conected with their transcriptional activity or with an increase in recombination variability and in the end to an increase in genetic variability of the species.

REFERENCES

Goodpasture, C. and Bloom, S.E. (1975). Expl Cell Res. 75, 304-306

Hsu, T.D., Spirito, S.E and Pardue, M.L., (1975). Chromosoma 53, 25-36.

Jones, R.N and Rees, H., (1982). B chromosomes, Academic Press, New York, p. 266.

Schmid, M.,(1978). Chromosoma 68, 131-148.

Schmid, M.,(1982). Chromosoma 87, 327-344.

Shmid, M., Sims, S.H., Haaf, T. and Macgregor, H.C., (1986) Chromosoma 94, 139-145.

Summer, H.T., (1972). Expl Cell Res., 75, 304-306.

Ullerich, F.H., (1967). Chromosoma 21, 345-368.

Cytogenetics of Amphibians and Reptiles
Advances in Life Sciences

THE KARYOTYPE, C-BANDS AND AG-NORS OF RANA SCHMACKERI

Wei Gang, Xu Ning and Li Dejun

Department of Biology, Zunyi Medical College, Guizhou Province, Popular Republic of China.

SUMMARY: *Rana schmackeri* has a diploid chromosome number of 26 with NF=52. There is a pronounced secondary constriction associated with the Ag-NORs located on 10q. Strongly positive C-bands are in a centric position on each pair of chromosomes. An interstitial C-band is located on 2q, 3q, 4p and 9q and a telomeric C-band is at the end of 5p.

INTRODUCTION

Comparative cytogenetic analyses among many species of the genus *Rana* have contributed to the understanding of karyotype evolution in the Anura (Morescalchi 1973, Schmid 1978).

There are 9 species of the stink frog group in China. Of them, studies on the karyotypes, C-bands and Ag-NORs of *R.grahami* and *R.margaratae* have been performed (Li Shushen et al. 1982, Wang Zishu and Wang Xizhong 1983, Liu Wanguo and Zan Ruigang 1984). In this paper both C-bands and Ag-NORs are used for analysing the karyotype. The results from C-banding and Ag-NOR banding in *R.schmackeri* can provide evidence for discussion about the evolution in the stink frog group of Rana genus.

MATERIAL AND METHODS.

Two males and one female of R.schmackeri were collected from Daozhen county, Guizhou, China, in May 1987. Both ends of the femur, tibiofibula, humerus were cut off and bone marrow cells were washed out with 0.4 M KCl for chromosome preparation by a centrifugal air-drying method. The samples were stained with 2% Giemsa. C-banding was performed following Sumner´s (1972). Ag-NOR banding was made by a developer Ag-NOR banding technique from Howell and Black (1980).

RESULTS

More than 50 metaphase spreads were observed and counted. R.schmackeri has a diploid chromosome number of 26 with 5 pairs of large (relative length greater than 9%) and 8 pairs of small (relative length lesser than 7%) chromosomes (Table 1). The secondary constriction (SC) can be found on 10q.

The results of C-banding and Ag-NORs banding are presented in figure 1. There is a strongly positive C-band in the centric position on each pair of chromosomes.An interstitial C-band is located on 2q, 3q, 4p and 9q.A telomeric C-band is at the end of 5q. In addition, the polymorphism of the telomeric C-band at the end of 9q between homologous chromosomes can be recognized,i.e. a telomeric band can only be found in one of the two homologues.There is a strong Ag-NOR associated with SC on 10q.

Table 1. Relative length (rl), centromeric index (CI) and centromere position (CP) in *Rana schmackeri*.

Chromosome no.	Rl	CI	CP
1	14.0+0.7	44.5+2.1	M
2	12.1+0.7	38.8+2.0	M
3	11.0+0.7	34.8+5.6	SM
4	10.9+0.6	41.7+7.3	M
5	9.3+0.5	46.1+3.9	M
6	6.7+0.3	46.2+2.8	M
7	5.9+0.4	30.9+6.1	SM
8	5.7+0.3	41.1+6.9	M
9	5.6+0.2	32.2+0.2	SM
10	5.1+0.3	44.4+4.3	M
11	4.9+0.4	43.2+3.1	M
12	4.4+0.5	42.5+3.1	M
13	4.2+0.4	44.5+4.1	M

Fig.1 The karyotype (a); C-banding (b) and Ag-NOR (c) of *R.schmackeri*.

DISCUSSION

1.Karyotype.

R.schmackeri has a karyotype of 2n=26 with 10 metacentric and 3 submetacentric chromosome pairs, and it´s closely related stink frog *R.grahami* has a similar karyotype, while the other stink frog *R.margaratae* has a karyotype of 2n=26 with 8 metacentric and 5 submetacentric chromosome pairs. The arm ratios of the chromosomes are useful in differentiating closely related species (Zhao et al. 1987). Moreover Morescalchi (1973) proposed that the more primitive species tend to have more submetacentric, subtelocentric and telocentric chromosomes. According to this point of view *R.margaratae* may be more primitive than *R.grahami* and *R.schmackeri*.

In the evolution of Amphibia, the rate of gene mutation is more rapid than karyotype mutation (Wilson et al. 1975). The level of gene mutation can be revealed by the SC number, and it has been speculated that more evolved species have a greater number of SC (Xiamao Zheng and Guanfu Wu 1989). The pronounced SC in the *Rana* species usually is located on 10q. *R.schmackeri* has only a pair of SC located on 10q; *R.grahami* has a pronounced SC located on 10q and 11 pairs of small SC on other chromosomes; while *R.margaratae* has a pronounced SC located on 11q and 10 pairs of additional variable small SC on some other chromosomes (Li Shushen et al. 1982; Wang Zishu and Wang Xizhong 1983; Liu Wanguo and Zan Ruigang 1984). So as to the number of SC *R.schmackeri* seems to be more primitive and *R.margaratae* more evolved.

NORs

R.schmackeri has only a large NORs pair on 10q; *R.margaratae* has also only a large NORs pair on 11q (Chen Wenyuan et al. 1983); while *R.grahami*, in addition to a large NORs pair on 10q, has 10 small additional NCRs on some other chromosomes. Hsu (1975) hypothesized that a karyotype with a single NOR is more primitive than those in which the NORs are distributed on several chromosomes. According to this hypothesis *R.grahami* is perhaps more evclved than the other two stink frogs as regards the NOR.

Table 2. Primitive and advanced karyological characters of the stink frogs.

Characters	*R. grahami*	*R. margaratae*	*R. schmackeri*
Submetacentric chromosomes	3 pairs	5 pairs	3 pairs
Number of SC	12 pairs	11 pairs	1 pair
Number of NORs	12 pairs	1 pair	1 pair
Interstitial C-bands	5 pairs	1 pair	4 pairs
Telomeric C-bands	3 pairs	26 pairs	2 pairs
No of primitive characters	0	4	2
No of advanced characters	5	1	3

3.C-bands.

Most species of *Rana* have many centromeric and telomeric C-bands, and few interstitial C-bands. All the three stink frogs have centromeric C-bands on each pair of chromosomes. R.margaratae has only one interstitial C-band (3q) and telomeric C-bands at the end of each chromosome arms. R.schmackeri has 4 interstitial C-bands (2q, 3q, 4p and 9q) and 2 pairs of telomeric C-bands (5q and 9q). R.grahami has 5 pairs of interstitial C-bands (3q, 4p, 6q, 10q and 11q) and 3 pairs of telomeric C-bands (5q, 6p and 7q). So R.margaratae might be more primitive than the other two stink frogs as regards the heterochromatin.

In summarizing the various karyological characters examined in the three stink frogs it can be speculated that R.margaratae is probably the karyologically most primitive species, R.grahami is the most evolved species, R.schmackeri occupying an intermediate position.

REFERENCES

Chen Wenyuan, Wang Zisu, Wang Xizhong, Yang Yuhua and Sun Qiling (1983). Zool. Res. 4, 82-88.
Howell, W.M. and Black, D.A. (1980) Experientia 36, 1014.
Hsu, T.C. (1975) Genetics 79, 137-150.
Li Shushen, Wang Yingxiang and li Chongyun (1982) Acta Genetica Sinica 9, 473-478 (In Chinese).
Liu Wanguo and Zan Ruigang (1984) Acta Genetica Sinica 11, 294-301. (In Chinese).
Morescalchi, A. (1973) In: Cytotaxonomy and Vertebrate Evolution (A.B. Chiarelli and E. Capanna Eds.) Academic press, London,New York, pp. 270-273.
Schmid, M. (1978). Chromosoma 68, 131-148.
Sumner, A.T. (1972) Exp. Cell Res. 75, 304-306.
Wang Zishu and Wang Xizhong (1983) Acta Herpetologica Sinica 2,1-6 (In Chinese).
Wilson,A.C., Bush, G.L., Case, S.M., and King. M.C. (1975) Proc. Natl. Acad. Sci. U.S.A. 72, 5061-5065.
Xiaomao Zheng and Guanfu Wu (1989) Chin. Herpetol. Res. 2, 37-45.
Zhao,E., Tan, A. and Wu G. (1987).Chin. Herpetol. Res. 1, 7-11.

PART 2. REPTILIA

Cytogenetics of Amphibians and Reptiles
Advances in Life Sciences

CHROMOSOMAL AND IMMUNOGENETIC DATA: A NEW PERSPECTIVE ON THE ORIGIN OF AUSTRALIA'S REPTILES

Max King

Museums of Arts and Sciences of the Northern Territory
G.P.O. Box 4646, Darwin, Australia 0801

SUMMARY: It is now clear that the often repeated theory that Australia was effectively devoid of endemic reptiles, and that it was colonized by multiple invasions from the north, is open to challenge. Comparative chromosomal studies coupled with microcomplement fixation analyses, suggests that the Gekkonidae, and Scincidae are Gondwanaland relics. Chromosome morphology also indicates that the Pygopodidae evolved from the Carphodactylini lineage subsequent to Australia's isolation. These data also suggest that the invasion and radiation of Asian and African faunal groups such as the Varanidae, Agamidae, Elapidae and Crocodilia has been an ongoing process following the contact of the Australian tectonic plate with Asia.

THE ORIGIN OF AUSTRALIA'S REPTILES: THE CURRENT VIEW

Hypotheses explaining the probable origins of the Australian reptile fauna abound. These range from the "uninhabited wastelands" approach adopted in many of the earlier studies (Darlington, 1957; Storr, 1964). To those which attempt to integrate continental drift theory into our understanding of the evolution of Australia's reptilian fauna (Cracraft, 1974; Tyler, 1979). All of these attempts have utilized essentially similar data bases which include the distribution of each family, the fossil record of that family and the degree of diversification of its members within and apart from Australia. However, there is substantial disagreement as to which reptile families arose in Australia when it was part of Gondwanaland and remain as relics, and which families arose in other parts of the world and radiated into Australia after its contact with the Asian tectonic plate (Table I). Perhaps this is not so surprising, for the Australian fossil record is very poor and in most cases fossils of modern reptiles are only dated back to the Miocene (Molnar, 1982), thus removing a source of most valuable information which is the very basis for determining the antiquity and distribution of taxa. However, there also appears to be a significant amount of resistance to the acceptance of Gondwanaland as a possible area for the origin of Australian reptiles (compare families listed by different authors in Table I).

The dichotomy of views is best seen if we compare Cogger and Heatwole (1981) who "suggest that most of the modern families in Australia's reptile fauna, with the exception of chelid tortoises, are post-Gondwanic immigrants from Asia" (pp 1336), with Cracraft (1974) who listed the Chelidae, Agamidae, Gekkonidae, Pygopodidae and Boidae (in part) as having a Gondwanaland origin. Tyler (1979) felt that the Chelidae and Gekkonidae were Gondwanaland relics and that the Pygopodidae arose from the Gekkonidae relatively recently, whereas the fossil crocodiles in Australia were of a Gondwanaland origin the extant crocodiles were of an Asian origin.

The main problem seems to be that we have reached an impasse as to what comparative morphological analyses can tell us about the zoogeography of Australian reptiles. With this in mind, the present study attempts to place a different evolutionary perspective on this reptile fauna using additional data sets. Since we now have chromosomal analyses completed on some 403 of the 693 Australian reptiles (see Table II), with a high level of coverage in each family, it is possible to provide a breakdown of the major karyotypic groups and determine their distribution and likely origin. Equally, with the significant amount of immunogenetic data now available, most of which uses the microcomplement fixation technique (MCF) to catalogue the evolution and rate of change of the albumin molecule, we can test the veracity of karyotypic groups, interpret their likely origins, and most significantly place a time scale on the origin of the

TABLE I. The origins of Australian reptiles proposed by Cracraft (1974), Tyler (1979), Cogger and Heatwole (1981).

Gondwanaland Origin	Laurasian Origin	Undefined Origins
Cracraft (1974)		
Chelidae	Carettochelydidae	Cheloniidae
Agamidae	Scincidae	Dermochelyidae
Gekkonidae	Varanidae	Typhlopidae
Pygopodidae	Boidae (in part)	Colubridae
Boidae (in part)	Elapidae	Hydrophiidae
		Crocodylidae
Tyler (1979)		
Chelidae	Varanidae	
Gekkonidae	Scincidae	
Pygopodidae	Agamidae	
Crocodylidae (fossil)	Elapidae	
	Colubridae	
	Acrochordidae	
	Typhlopidae	
	Boidae	
Cogger and Heatwole (1981)		
Chelidae	Pygopodidae	Cheloniidae
	Hydrophiidae	Dermochelyidae
	Gekkonidae	Laticaudidae
	Scincidae	Acrochordidae
	Agamidae	Colubridae (in part)
	Elapidae	
	Crocodylidae	
	Boidae	
	Typhlopidae	
	Varanidae (?)	
	Carettochelydidae	
	Colubridae (in part)	

different higher taxa and on the differentiation of species. In short, when these data sets are used in tandem, they provide a new and corroborative perspective on the evolution of Australia's reptile fauna.

WHAT CHROMOSOMAL AND MCF DATA TELL US ABOUT THE ORIGINS OF AUSTRALIA'S REPTILES

In this overview of the origin of Australia's reptiles, most of the information presented will consider the suborder Squamata; the lizards. This is because of the irregular coverage of chromosomal and MCF data on the remaining reptilian taxa. Two notable exceptions to this are the chromosomal, electrophoretic and MCF analyses of Australian elapid snakes (Mengden, 1985a; Schwaner et. al., 1985), and similar studies on the crocodilians (Cohen and Gans, 1970; Densmore, 1983). These investigations have determined that the extant crocodilians are very recent invaders of northern Australia being part of the post Pliocene *Crocodylus* radiation (1 million years before present (MYBP)). Nevertheless, the Crocodilians in toto are an ancient group first diverging into three major lineages (alligators, crocodiles and gavials) some 70 MYBP in the Cretaceous/Tertiary. The precise area of origin of these organisms remains unknown. The major radiation of Australian elapid snakes was dated at 12 MYBP. The Hydrophiidae and Laticaudidae are now known to be monophyletic with the Elapidae. The Hydrophiids diverged from the *Notechis* lineage of the Elapidae some 5 MYBP. Clearly, the origins of these taxa can not be directly linked with Gondwanaland for the separation of Australian elapids from their Asian and African counterparts only occurred some 20 MYBP (Schwaner et. al., 1985).

S.O. SQUAMATA

1. The Gekkonidae

This most diverse and ancient group of reptiles has a circumglobal distribution. The four subfamilies, Eublepharinae

TABLE II. The number of species of Australian reptiles and those analysed chromosomally.

Family	Number of species	Number of species karyotyped	References
Class Reptilia			
Order Squamata			
Suborder Sauria			
Gekkonidae	93	72	King, 1987a, b and c.
Pygopodidae	30	15	King unpublished data
Varanidae	26	17	King & King, 1975; unpublished data
Scincidae	292	134	King, 1973; Donnellan, 1985
Agamidae	63	22	Witten, 1983
	504	260	
Suborder Serpentes			
Typhlopidae	31	6	Mengden pers. comm.
Acrochordidae	2	2	Mengden, 1983
Boidae	13	13	Mengden pers. comm.
Colubridae	10	8	Mengden, 1983; pers. comm.
Elapidae	76	72	Mengden, 1985a, and b
Laticaudidae	2	2	Mengden, 1983
Hydrophiidae	30	19	Mengden pers. comm.
	164	122	
Order Crocodilia			
Crocodilidae	2	2	King et. al., 1985
Order Testudines			
Cheloniidae	5	3	Bickham & Carr, 1983
Dermochelyidae	1	-	-
Carettochelydidae	1	1	Bickham et. al., 1983
Chelidae	16	15	Bull & Legler, 1980
	23	19	
Grand Total	693	403	

(5 genera), Sphaerodactylinae (5 genera), Diplodactylinae (14 genera) and Gekkoninae (58 genera) are found in regional clusters throughout this distribution. Thus, the Eublepharinae are found in North America, Mexico, east and west Africa, the Middle East and southern Asia where they occur in apparently isolated clusters. The Sphaerodactylinae are restricted in their distribution to northern South America and Central America. The Diplodactylinae occur in Australia, New Zealand, New Caledonia and the Loyalty islands. The Gekkoninae are found throughout the world excluding the icecaps, although it should be noted that many of the genera are restricted to Africa, the middle east, Australia and South America and it is only some genera which Kluge (1967) termed "modern expanding dominants" which have truly world wide distributions.

The origin of the Gekkonid subfamilies is most difficult to determine if one simply looks at the present day world distribution. However, if the continents and islands are reorganized into a Gondwanaland supercontinent (see King, 1987a, 1987b), we see that all subfamilies are found on the one land mass and that certain of the additional distributions in Asia and other regions can be attributed to subsequent radiations following the breakups of Gondwanaland and the secondary contact of Africa and India with Europe and Asia respectively. While these distributional factors and the antiquity of the complex (upper Jurassic) suggest a Gondwanaland origin, what can be learned from chromosomal analyses?

In terms of the Eublepharinae and Sphaerodactylinae, too few species have been analysed chromosomally to make any pertinent comment on the evolutionary signficance of the differences. The four Eublepharine species analysed show considerable variation in number and form, although it should be noted that acrocentric karyotypes with 2n=38 and 2n=32 are present (DeSmet, 1981; Matthey, 1933). In the Sphaerodactylinae, six species from the genus <u>Gonatodes</u> have been analysed by McBee et. al., (1984, 1987), some of these have acrocentric karyotypes with 2n=32, but others have numbers ranging from 2n=22 to 40.

In the Gekkoninae, some 74 species have been analysed chromosomally, and while this is a very unevenly distributed sample from the 58 known genera, a series of features are apparent:

1. Eight distinct putative ancestral karyomorphs are present each of which is acrocentric or telocentric. Clusters of species possess each of these karyomorphs, and chromosome numbers include 2n=32, 34, 36, 38, 40, 42, 44, 46.
2. Numerous fusions, inversions, additions and tandem fusions have occurred within each of these categories. Species with these derived karyotypes can be assigned to each of the ancestral karyomorph groups.
3. These data suggest that the Gekkoninae are a polyphyletic assemblage, and there is some agreement between the chromosomal groupings of King (1987a) and morphogroups defined by Russell (1972) and Kluge (1987).

Joger (1985) made an immunological comparison of the serum albumins of some 35 Gekkonine species derived from ten African genera and outgroups from southern Asia and South America. The analyses were one-way tests using the antisera of three <u>Tarentola</u> species and one <u>Pachydactylus</u>, against the whole serum of test species. Immunological distances were calculated from the level of precipitin.

One-way tests suffer from the difficulty of resolving different rates of evolution and are thus not as powerful as reciprocal tests using the antisera for each test species. Nevertheless, the data are informative and in terms of major groups are credible. Thus, South American and Asian genera are clustered at the greatest distance (90 MYBP). African genera are grouped into three major clades: a southern African radiation (60 MYBP) including <u>Pachydactylus</u>, <u>Palmatogecko</u> and <u>Rhotropus</u>. These agree with Russell's (1972) <u>Pachydactylus</u> group, second, an Afro-Madagascan group (70 MYBP) including

Phelsuma, Homopholis and Lygodactylus. The age of this group equates with the estimated date for the separation of Madagascar from Africa (60 MYBP) (Joger, 1985). The third is a north African radiation including Tarentola, Ptyodactylus and Stenodactylus. Neither of the latter groups equate particularly well with Russell's (1972) clustering. These data do support the concept that the Gekkoninae have a Gondwanaland origin

In direct contrast, the Diplodactylinae are a monophyletic group based on the evolution of a 2n=38 karyotype (King 1987a and b). Here, a 2n=38 all acrocentric karyotype with minute short arms is common to members from both the Diplodactylini and Carphodactylini. Clusters of low vagility terrestrial species within the tribe Diplodactylini have established fusion differences between species, and these may be associated with speciation (King, 1977). In contrast, arboreal species from the D. ciliaris species group have established inversion differences on six chromosomes and this karyomorph characterizes that group. Most significantly, species from the genus Oedura have pericentric inversion differences established on every chromosome in the complement, thus changing the ancestral 2n=38 karyotype to a biarmed karyotype in which all chromosomes have distinct short arms i.e. acrocentric, submetacentric and metacentric elements are present (Fig. 1a). This highly derived karyomorph is shared with a number of genera in the Carphodactylini: Rhacodactylus and Bavayia from New Caledonia, Hoplodactylus from New Zealand and Pseudothecadactylus and Carphodactylus from Australia (Fig. 1b). Chromosomes from these complements are not only similar in their biarmed appearance, but this similarity extends to chromosome arm lengths and arm ratios for particular chromosomes. Certain differences have been established between individual species, but the overall appearance of homology between genera is most apparent.

These data, suggest that the Diplodactylini are a polyphyletic assemblage and the Oedura should not be regarded

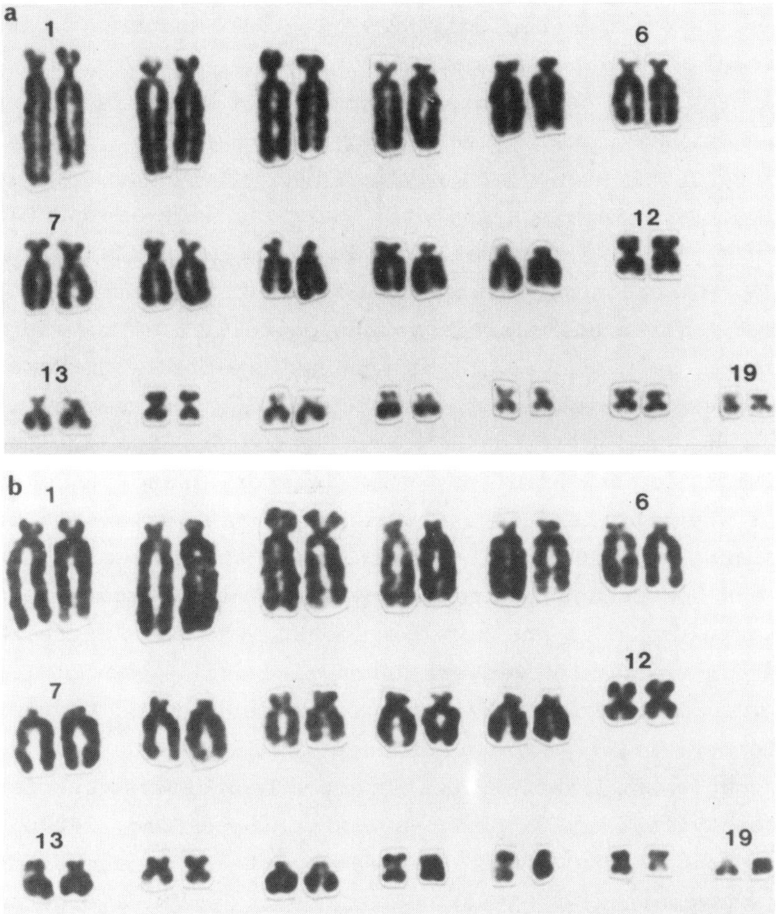

Figure 1a. The karyotype of Oedura marmorata which has the highly derived 2n=38 biarmed karyomorph. 1b. The karyotype of Hoplodactylus maculatus which also possesses the 2n=38 biarmed karyomorph. Compare the chromosome size, centromere position and arm lengths for each species.

as a member of the Diplodactylini but a member of the Carphodactylini (contra Kluge, 1967, 1987). Moreover, since the uniarmed 2n=38 acrocentic karyomorph is ancestral then, in the absence of convergence, the genera Oedura, Carphodactylus, Pseudothecadactylus, Rhacodactylus, Bavayia and Hoplodactylus are monophyletic (compare figs 1a and b). They differ by 19 fixed pericentric inversion differences (i.e. 19 characters), from the ancestral acrocentric karyomorph. These animals are morphologically most similar, being very large rock dwelling or arboreal species. The most reasonable explanation for the evolution of this complex is that this derived biarmed karyomorph attained its unusual distribution in New Zealand, New Caledonia and much of Australia when the land masses were in contact as a part of Gondwanaland some 80 million years BP. At that time New Zealand and New Caledonia broke away from the supercontinent and began to drift to the north and east, whereas Australia remained in contact with western Antarctica until some 55 MYBP when it started to drift to the north (Colbert, 1973; Veevers and McElhinny, 1976; Audley-Charles et. al., 1981).

Microcomplement fixation data obtained from several species of Diplodactylinae analysed by Baverstock (see King, 1987b), indicate that not only is the separation of the uniarmed and biarmed karyomorphs at least 66 MYBP but that divergence between genera retaining the acrocentric ancestral karyotype is also of this age. This difference extends to genera within the Diplodactylini and to differences between Diplodactylini and Carphodactylini. Even relatively closely related species from the same genera show twenty million years separation. Clearly, the immunological distances obtained from these gekkos indicate that the Gekkonidae were present in Gondwanaland before the breakup of that supercontinent.

2. The Pygopodidae

This endemic Australian lizard family comprising some 35 species remains as one of the enigmas of the reptile world.

They are distributed throughout the Australian mainland, excluding Tasmania. A single species, Lialis burtonis, is found in New Guinea.

Published chromosomal data are restricted to the chromosome number of Delma fraseri 2n=38 (Gorman, 1973) and the karyotype of Lialis burtonis 2n=33 X_1 X_2 Y (Gorman and Gress, 1970). A

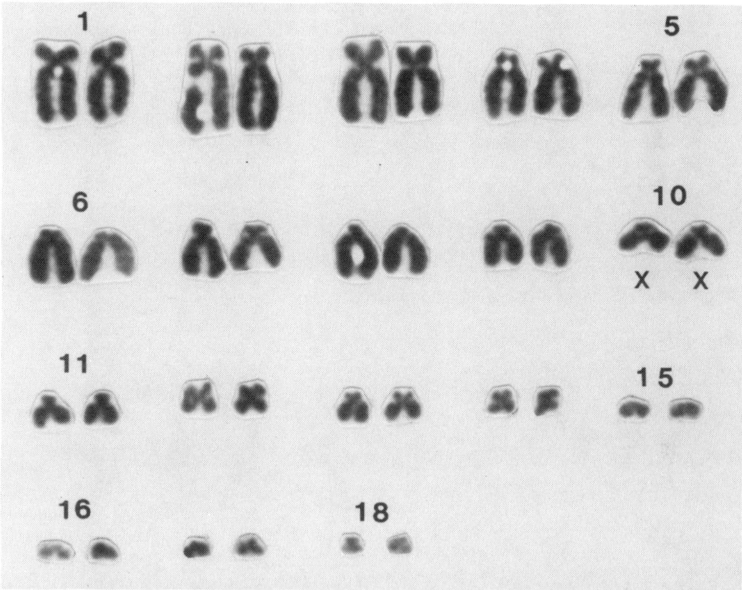

Figure 2. The karyotype of Delma inornata female with 2n=36. Note the homomorphic X chromosomes and compare to Fig. 3.

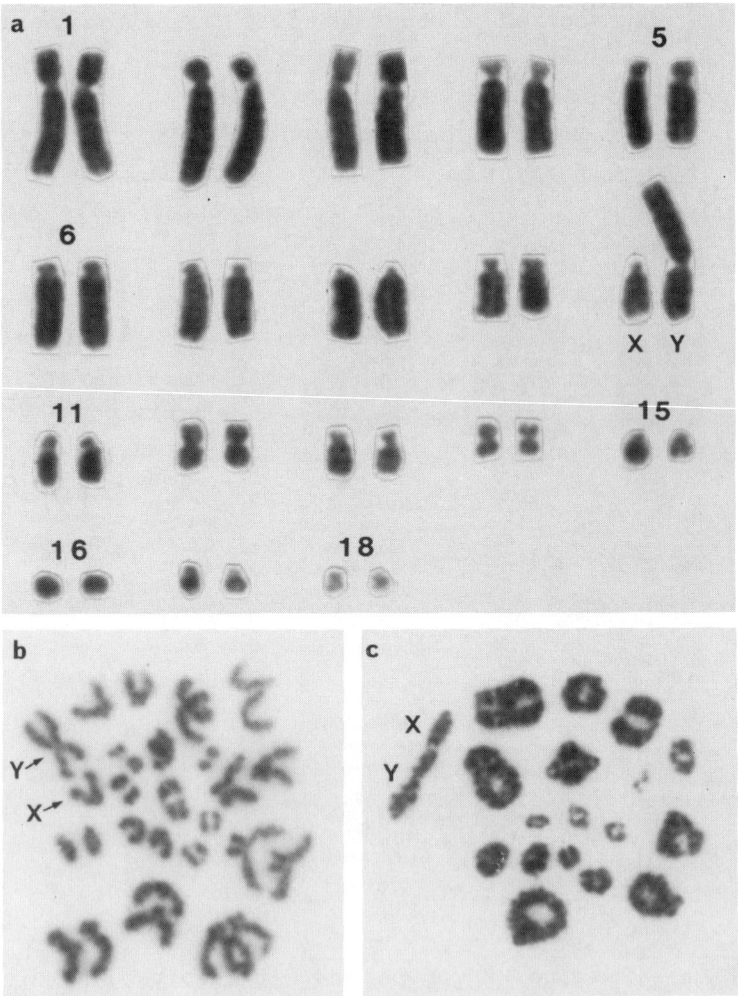

Figure 3a. The Karyotype of <u>Delma</u> <u>inornata</u> male 2n=36. Note heteromorphic pair 10 XY. b. meiotic anaphase cell showing X and Y elements separating (colchicine treated). c. Meiotic metaphase cell, note the XY bivalent.

survey of 15 Australian species from the genera Pygopus, Ophidiocephalus, Aprasia and Delma reveals a complex of species with chromosome numbers ranging from 2n=34 to 2n=42. The majority of species have sex chromosome systems based on male heterogamety. These include simple XY heteromorphisms to more complex systems involving multiple sex chromosomes (Gorman and Gress, 1970; King, unpublished data).

The karyotype of Delma inornata shown in Figs 2 and 3 has a biarmed 2n=36 format in male and female cells and a substantial heteromorphism in chromosome pair 10 in males. The Y chromosome is a large submetacentric which appears to have been derived from the acrocentric X by the addition of a chromosome arm. This does not appear to be heterochromatic. Other species from the genus Delma have 2n=38 biarmed complements as do species from the genera Pygopus and Ophidiocephalus (King, unpublished, data).

The chromosome morphology of the basic 2n=38 biarmed karyomorph is most revealing for many of the chromosomes have distinct short arms and are acrocentric, submetacentric, or metacentric. They share a striking resemblance to the biarmed 2n=38 karyotype found in many members of the Carphodactylini Compare Figures 1, 2 and 3). This similarity extends to arm length and chromosome lengths for many members of the complement. It would appear that this is too much of a coincidence to be attributed to convergence in chromosome morphology. The most reasonable explanation is that the Pygopodidae share the derived karyomorph found in certain groups from the Carphodactylini. This is to say that the Pygopodid lineage evolved from the biarmed lineage of the Carphodactylini. It is noteworthy that this view is significantly different to that of Kluge (1987) who argued that the Diplodactylini genera Crenadactylus, Diplodactylus, Oedura and Rhynchoedura share several cranial modifications found in the Pygopodidae. Indeed, Kluge hypothesized that the Australian - New Zealand separation would be consistent if pygopodids were the sister-group to the Crenadactylus +

Diplodactylus + Oedura + Rhynchoedura lineages of Kluge (1967). However, chromosomal data indicate that these genera are a polyphyletic assemblage.

The restricted distribution of the Pygopodidae suggests that this family arose in and radiated throughout Australia subsequent to the separation of Australia from Gondwanaland.

3. Agamidae

The lizards of the Agamidae are essentially an old world family distributed throughout Africa, south eastern Europe, India, the Middle East, South Eastern Asia and Australia. Wermuth (1967) recognized 291 species in 34 genera and many more species have been identified since that date.

Chromosomal studies on the Australian Agamidae are limited to the analysis of some 22 of the 66 species using standard chromosomal analysis (Witten, 1983; King, 1985). The vast majority of species appear to have been derived from a single radiation for species from the genera Amphibolurus, Diporiphora, Chlamydosaurus, Lophognathus, Ctenophorus, Pogona and Tympanocryptis all share a 2n=32 biarmed karyotype subdivided into six pairs of large metacentric or submetacentric chromosomes and 10 pairs of small metacentric microchromosomes (Fig. 4).

A second karyomorph was observed in Gonocephalus spinipes and Physignathus lesueurii where a chromosome number of 2n=36 was observed. This karyomorph consisted of 6 pairs of large metacentric and submetacentric elements and 12 pairs of metacentric microchromosomes (Witten, 1983). It is noteworthy that the 2n=40 karyotype reported for Lophognathus gilberti by Witten (1983) which possessed numerous telocentric elements was not observed in a subsequent examination by King (1985). Although specimens originated from a different area the karyotype was the standard and uniform 2n=32 complement. In what may be an analogous situation, Ota (1987) reported extensive differentiation of the Taiwanese agamid Japalara swinhonis where populations with 2n=36, 40 and 46 were encountered.

Witten (1982) suggested that the Australian Agamids could be divided into two groups on the basis of their osteology and morphology, namely, a small group of Asian derived genera including Chelosania, Physignathus and Gonocephalus, and a major endemic Australia radiation which includes Amphibolurus, Ctenophorus, Chlamydosaurus, Diporiphora, Lophognathus,

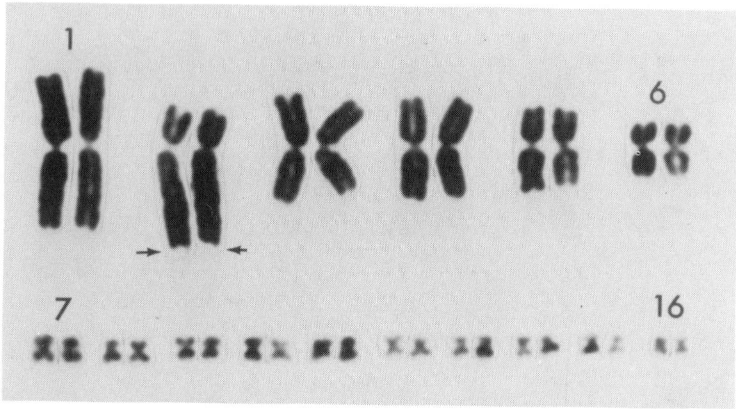

Figure 4. Typical 2n=32 Agamid metacentric karyotype (Pogona vitticeps). Note 6 pairs of larger chromosomes and 10 pairs of microchromosomes.

Pogona, Tympanocryptis, Caimanops and Moloch. This dichotomy is supported in so far as the chromosome morphology of the putative Asian derived species have a 2n=36 karyomorph, whereas the endemic Australian forms share the 2n=32 karyomorph.

If we examine the published chromosomal information on the Agamids obtained from other regions we see a remarkably different picture. Olmo (1986), Peccinini-Seale (1981) and Gorman (1973) list a total of some 72 species of Agamids examined chromosomally. Solleder and Schmid (1988) analysed a further 6 species and reexamined three previously karyotyped forms with a variety of banding techniques. Chromosomes numbers varied extensively between species ranging from a low of 2n=28 to a high of 2n=50. Most of these differences could be attributed to extensive fissioning of the lower numbered metacentric karyotypes or fusing of the acrocentric chromosomes. It is noteworthy that the 2n=36 karyotype, which Gorman (1973) argued was the ancestral karyomorph in lizards, is relatively uncommon in Agamids (an alternative ancestral karyomorph is provided by King, 1981).

It may be significant that the Australian Agamids appear to be a most karyotypically stable group. This suggests that a single radiation possessing the 2n=32 karyomorph diversified throughout Australia without any gross chromosome changes being associated with speciation. Morphological differences between species with the same karyomorph are enormous: ranging from the typical *Ctenophorus pictus* to the bizarre *Moloch horridus*. It is possible that the 2n=36 karyomorph group represents a subsequent invasion of Australian by Asian Agamids as Witten (1982) hypothesized, although they may equally well have been secondarily derived from the major radiation of endemic forms.

Baverstock and Donnellan (in press) used the microcomplement fixation technique to test the relationships between Asian, African and Australian Agamidae. Antisera were raised against albumin from six species from the genera *Ctenophorus*, *Pogona*, *Lophognathus*, *Moloch*, *Gonocephalus* and *Physignathus* and were tested in a reciprocal matrix. One way tests were made against the Australian genera *Tympanocryptis*, *Chlamydosaurus* and *Diporiphora* (3 species); New Guinean *Gonocephalus* (1 species); Asian *Gonocephalus* (1 species) and

Calotes (1 species) and African Agama (1 species). Dipsosaurus and Iguana were used as outgroups.

The data indicated that there has been a major endemic Australian agamid radiation and that this has been relatively recent i.e. less than 20 MYBP (Miocene and Pliocene). Australian species from the genera Physignathus and Gonocephalus, which were postulated as being of Asian origin, were in reality a part of the endemic Australian radiation. Australian Gonocephalus are closer to the other Australian genera, including Physignathus, than they are to Asian Gonocephalus. The divergence of Physignathus and Australian Gonocephalus from other Australian genera occurred in the mid Miocene.

Most surprisingly, all Australian agamids examined were closer to African Agama then they were to any Asian species. Thus, Baverstock and Donnellan (in press) attributed similarities between Asian and Australian Agamids to be a result of morphological convergence. The time of divergence of Asian and Australian stock was estimated at 60 MYBP, whereas, the African/Australian divergence was estimated at 40 MYBP. These data clearly suggest that the African/Australian connection and the age of the radiation of endemic Australian Agamids (20 MYBP), is far too recent to be attributed to a Gondwanaland origin. While, it could be argued that the divergence of extant Australian Agamids is 20 MYBP, and that older lineages are extinct, this would be highly unlikely. The uniformity of chromosomal morphology in endemic Agamids (2n=32) corroborates the MCF hypothesis of a single major radiation, followed by the subsequent divergence of Australian Gonocephalus and Physignathus (2n=34).

4. Scincidae

This widespread family of lizards with over 600 species is distributed throughout the tropical and temperate world. Nevertheless, by far the greatest species density occurs within Australia where some 292 species are now recognized.

Cogger and Heatwole (1981) argue that the Scincidae, while having a considerable history within Australia, are the product of multiple entries into that continent. That is, they have been resident in Australia since the Tertiary and have undergone a series of subsequent major adaptive radiations. This view is largely supported by Tyler (1979) who argues that skinks are adept at dispersal over sea water and thus the precise timing of their arrival in Australia is unknown, although they undoubtedly colonized Australia. Cracraft (1974) also argued that the Scincidae dispersed to Australia from South East Asia and that they had a Laurasian origin. The new world forms appeared to be derived from Eurasian ancestors since the family is known from the late Cretaceous in North America.

While it is clear that the Scincidae are well adept at rafting, and may have attained much of their distribution particularly throughout Melanesia and Polynesia by this process, the fact remains that the absence of an effective fossil record from Australia has once again mitigated against any hypothesized Gondwanaland origin. This view has been promulgated despite the enormous diversity and density of species and the high level of endemism within Australia.

Very few species from the Scincidae have been analysed chromosomally from other than Australian sources (Gorman, 1973; Olmo, 1986). Diploid numbers range from 2n=24 to 32 and the karyotype is divided into 4 pairs of large metacentric elements along with three pairs of smaller chromosomes gradually diminishing in size to the smallest microchromosomes (Fig. 5).

Australian species were subdivided into three recognizeable clusters by King (1973) on the basis of chromosome number and morphology. Species from the genera *Egernia* and *Tiliqua* had 2n=32, whereas those from *Leiolopisma*, *Sphenomorphus* and *Ctenotus* had 2n=30. Moreover, *Ctenotus* and *Sphenomorphus* could be separated from *Leiolopisma* on the larger size of chromosome pairs 5 and 6 in the former cluster.

Donnellan (1986) extended this study of the Australian Scincidae with his thorough examination of 133 species. He too found three karyomorph groups; the Egernia group (Egernia, Tiliqua); Eugongylus group (Carlia, Cryptoblepharus, Emoia,

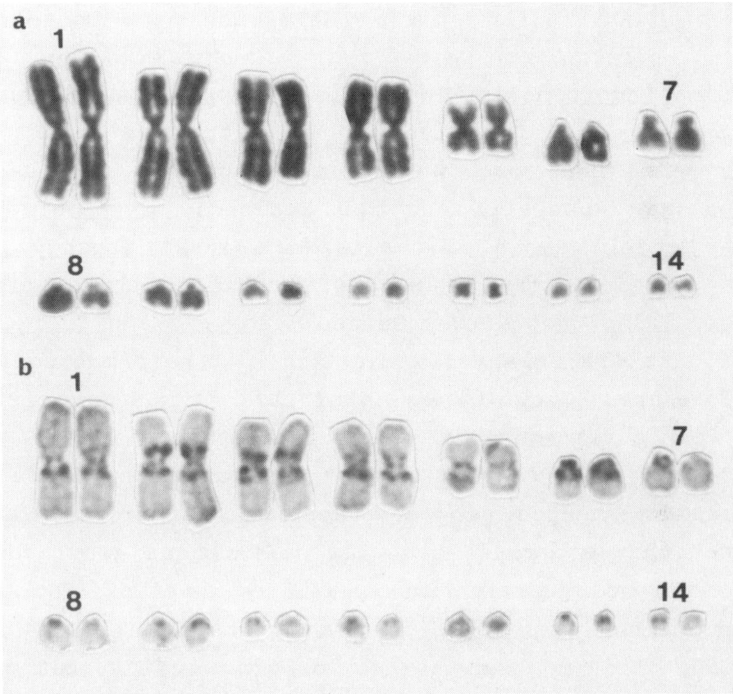

Figure 5a. The karyotype of Chalcides chalcides from Portugal. This typically scincid karyotype has 2n=28 chromosomes. 5b. The C-banded complement of this species.

Eugongylus, Lampropholis, Leiolopisma, Menetia, Morethia, Nannoscincus and Proablepharus); and Sphenomorphus group (Anomalopus Calyptotis, Ctenotus, Eremiascincus, Hemiergis, Lerista, Lipinia, Lobulia, Notoscincus, Prasinohaema, Saiphos, Sphenomorphus and Tropidophorus), separated by chromosome number and morphology.

A microcomplement fixation analysis has been made on 25 species of Australian skinks (Baverstock and Donnellan in press). Antisera were raised from 10 Australian species from the genera Lampropholis, Morethia, Leiolopisma and Egernia and reciprocally tested (although the matrix is incomplete). These were then one way tested against species from Leiolopisma (Australia and New Zealand), Cryptoblepharus, Carlia, Menetia, Emoia (Australia/New Guinea), Eugongylus (New Guinea), Sphenomorphus, Mabuya (Indonesia), Lamprolepis (New Guinea), Tribolonotus (New Guinea), Egernia and Tiliqua.

The data are relatively clearcut in their message. The Scincidae are a very old family of lizards in which the three Australian chromosomal lineages diverged some 60 MYBP. That is, the MCF data confirm the existence of the three basic chromosomal lineages. Indeed, such a time scale, coupled with the fact that enormous molecular differences are recognized between morphologically closely related species (i.e. 24 MYBP between species of Leiolopisma), suggests that the Australian Scincidae had a Gondwanaland origin. This pattern of diversification appears to be analogous to that seen in the Gekkonidae and contrasts with the Agamids.

It is noteworthy that the New Zealand species of Leiolopisma which was analysed appeared to be an unambiguous member of the Australian Eugongylus group. However, its divergence time from the nearest Australian relatives of 20 MYBP indicates that the New Zealand distribution could not be the product of a Gondwanaland origin, but may well be the result of a more recent invasion of New Zealand presumably by rafting (Baverstock and Donnellan in press).

5. The Varanidae

A single genus Varanus which is comprised of some 35 species is the sole taxa of these most unusual reptiles. This old world family is distributed throughout Africa, the middle east, southern USSR and China, India, South East Asia, and Australia. There are some 26 species found in Australia, where a substantial radiation of numerous lineages has occurred.

Whereas members of the Varanidae may be traced back to the upper Cretaceous (70 MYBP) in Asia and North America, their known fossil record in Australia is restricted to the giant extinct varanid Megalania prisca found in Miocene deposits (15 MYBP) (Stirton et. al., 1961).

On the basis of distribution and the fossil record Tyler (1979), Cracraft (1974) and Cogger and Heatwole (1981) argue that the Varanidae arose in Asia and radiated into Australia on a number of occasions.

King and King (1975) made a chromosomal analysis of some 16 of the 32 Varanid species. All species had similar 2n=40 karyotypes. However, we were able to define six independent karyomorph groups on the basis of shared derived and ancestral chromosome markers (Fig. 6a). These differences involved fixed multiple pericentric inversion of the medium sized group of chromosomes; the presence of acrocentric or metacentric microchromosomes; or the presence or absence of a short arm on chromosome pair 5. Clusters of species could be placed in each of the karyomorph groups. This karyotypic analysis has now been expanded to include some 27 species and subspecies which are classified as follows:

Group A. V. gouldii flavirufus, V. giganteus, V. gouldii gouldii, V. mertensi, V. rosenbergi, V. spenceri.

Group B. V. varius, V. komodoensis, V. indicus.

Group C. V. nebulosus, V. bengalensis, V. flavescens, V. karlschmidti, V. rudicollis, V. salvator.

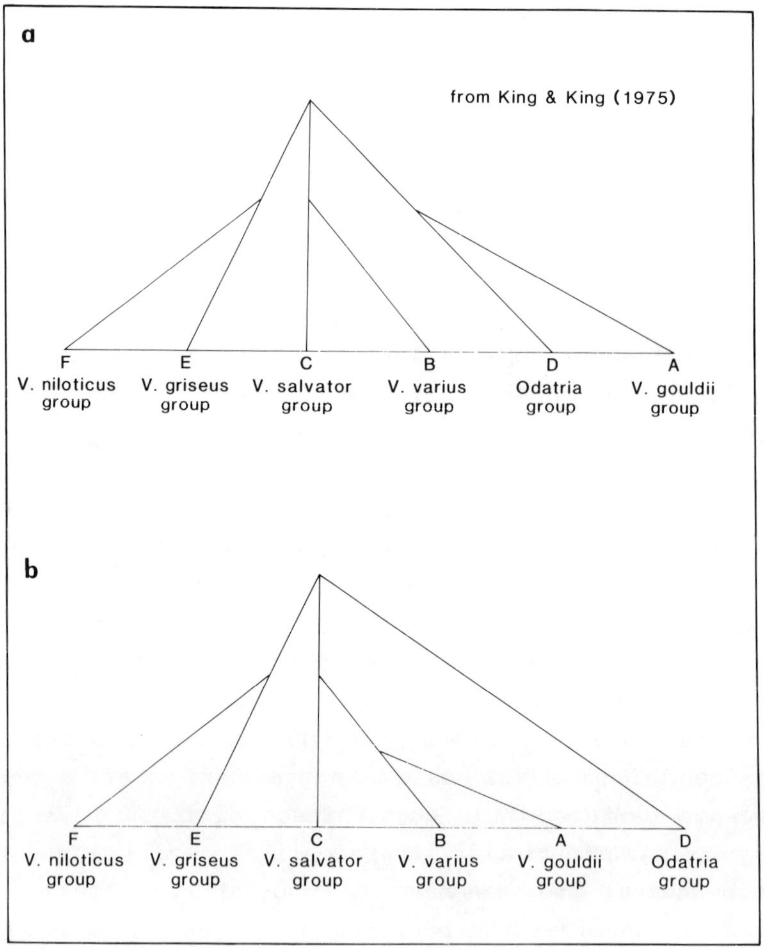

Figure. 6a. Phylogenetic relationships of the Varanidae based on karyotpic data from King and King (1975). Letters A to E refer to karyomorph groups. b. A revised phylogeny of the Varanidae using chromosomal and MCF data.

Group D. V. acanthurus, V. gilleni, V. kingorum, V. baritji, V. mitchelli, V. scalaris, V. tristis, V. semiremex, V. storri.

Group E. V. griseus

Group F. V. exanthematicus, V. niloticus.

An MCF analysis of 32 of the 35 Varanus species has recently been completed by Baverstock, King and King (unpublished). In this analysis antisera were raised to each of 16 species and reciprocal tests were made on these. One way tests were made on a further 16 species. Most significantly, there was agreement between karyomorph groups and MCF clusters in all cases but one. The MCF data indicate that Varanus indicus was a member of the Asian assemblage rather than being grouped with V. varius and V. komodoensis. This suggests that the subtle chromosomal differences in the microchromosomes which distinguished this species from the other Asian forms resulted from a convergence in chromosome morphology.

The results of this analysis indicate that the major lineages within the extant Varanus diverged no greater than 40 MYBP. Such a finding places the source of the Australian members of this taxa within the context of an Asian rather than a Gondwanaland origin.

MCF data define 3 clades arising from an unresolved polytomy and a fourth clade arising from the Asian lineage. The six karyomorph groups resolve clusters of species within each of these clades. A revised phylogeny based on combined MCF and chromosomal data for the evolution of this group is shown in Fig. 6b. While supporting the possibility of a multiple invasion hypothesis, the number of invasions into Australia from Asia has been increased by one. That is, an initial invasion and radiation of Odatria occurred. Subsequently, an invasion of the V. varius Group B species, was established and the Group A species arose from this lineage and radiated throughout Australia. Most recently the northern coastline of Australia has been colonized by V. indicus. The remaining radiations of middle eastern and African forms are also supported by the MCF data.

The chief difference between the phylogeny produced by King and King (1975) and the current view is that the V. gouldii complex (Group A) did not arise from the Odatria (Group D) lineage, as originally proposed, but arose from the Group B lineage, which itself was a derivative of the Group C lineage, both of which are Asian lineages. This is a less parsimonius relationship on chromosomal grounds, but undoubtedly has more morphological acceptance (compare Figs 6a and b).

All evidence points to the fact that the highly mobile varanid lizards have invaded diverse regions throughout their evolutionary history. Chief amongst these has been the multiple invasion of Australia by Asian species.

CONLUDING REMARKS

Because of the absence of fossilized reptiles older than the Miocene many of the assumptions on the origins of the Australian reptile fauna have been based on little more than intuition and past histories of those taxa in other regions (compare Tables II and III). This study has determined three distinct evolutionary patterns:

1. The use of both chromosomal and immunogenetic analyses in tandem have enhanced the possibility of determining group relationships within higher taxonomic categories. In those studies discussed here, where comparable groups were analysed, chromosomal and immunogenetic clusters were congruent in their findings. That is, higher taxonomic categories defined by chromosomal data were supported by MCF data. This applies to the Varanidae, Scincidae, Agamidae, and Elapidae. Too few species from the Gekkonidae and no Pygopodids where examined with MCF analyses to determine group relationships, although in the Gekkonidae and the other taxa mentioned above, the date of divergence of the extant taxa could be ascertained.

TABLE III. The age and origin of extant Australian reptiles based on the rate of albumin evolution. The approximate age assumes that the rate of albumin evolution is the same as that obtaining in other vertebrates.

	Age of radiation in Australia ($x10^{-6}$ years B.P.)	Age of separation from populations in other regions ($x10^{-6}$ years B.P.)	Probable area of origin of Australian taxa	Reference
S.O. Squamata				
Gekkonidae (Diplodactylinae)	>66	>66	Gondwanaland	Baverstock in King 1987b
Agamidae	20	40 African 60 Asian	Asia	Baverstock and Donnellan (in press)
Scincidae	60	20 New Zealand	Gondwanaland	Baverstock and Donnellan (in press)
Varanidae	40	40 African	Asia	Baverstock, King and King (in manuscript)
S.O. Serpentes				
Elapidae	12	20 Asian, African	Asia	Schwaner et.al., 1985
Laticaudidae	5	5 Australia	Asia	Schwaner et. al., 1985
Hydrophiidae	5	5 Australia	Asia	Schwaner et. al., 1985
O. Crocodilia	1	70	Asia	Densmore, 1983

2. There seems to be little doubt that the most common origin of Australian reptiles has been by the invasion of Australia by Asian or African lineages and this applies to the Agamidae, Crocodilia, Varanidae (3 invasions), and Elapidae. Equally, there is little doubt that the Australian Scincidae and Gekkonidae are relics of a Gondwanaland origin. Chromosomal data suggest that the Pygopodidae arose from the Carphodactylini lineage subsequent to the isolation of Australia from Gondwanaland.

3. Perhaps one of the most significant findings of this comparative chromosomal and immunogenetic approach is the great divergence between families in the rate of chromosomal evolution over known time periods. Thus, the Agamidae, Varandiae, Scincidae and the Gekkonidae (in part) have been most conservative in chromosome form throughout their evolutionary history. In contrast, the Crocodilians, Pygopodidae, Gekkonidae (in part) and Elapidae (in part) have shown high levels of chromosomal repatterning, much of which has been associated with species diversity.

ACKNOWLEDGMENTS

Thanks to Dr. Peter Baverstock and Dr. Steve Donnellan for providing me with access to their unpublished data and for their many helpful comments.

REFERENCES

Audley-Charles, M.G., Harley, A. M. and Smith, A.G. (1981) In: Wallace's line and Plate Tectonics (T.C. Whitmore Ed.), Clarendon Press, Oxford, pp. 9-23.

Baverstock, P.R. and Donnellan S. (in press). Mem. Queensland Mus.

Bickham, J.W., Bull, J.J. and Legler, J.M. (1983) Cytologia 48: 177-183.

Bickham, J.W., and Carr, J.L. (1983) Copeia, 918-932.

Bull, J.J. and Legler, J.M. (1980) Canad. J. Zool. 58: 828-841.

Cogger, H.G. and Heatwole, H. (1983) In: Ecological Biogeography of Australia (A. Keast Ed),. Dr W. Junk, The Hague, pp. 1333-1373.

Cohen, M.M. and Gans, C. (1970). Cytogenetics 9: 81-105.

Colbert, E.H. (1973) In: Implications of Continental drift to Earth Sciences (D.H. Tarling and S.K. Runcorn Eds), Academic Press, London. Vol. 1 pp. 395-412.

Cracraft, J. (1974) Ann. Rev. Ecol. Syst. 5, 215-262.

Darlington, P.J. (1957) Zoogeography: The geographical distribution of animals. Wiley: New York.

Densmore, L.D. (1983). In: Evolutionary Biology 16 (M.K. Hecht, B. Wallace and G.T. Prance Eds), Plenum pp. 397-465.

DeSmet, W.H.O. (1981) Acta. Zool. Pathol. Antverp. 76, 53-72.

Donnellan, S. (1985) Sex Chromosome Evolution in Lygosomine skinks. PhD thesis. Macquarie University.

Gorman, G.C. (1973) In: Cytotaxonomy and Vertebrate Evolution. (A.B. Chiarelli and E. Capanna Eds). Academic Press, London, pp. 349-424.

Gorman, G.C. and F. Gress (1970) Experientia 26: 206-207.

Joger, U. (1985) In: Proceedings of the International Symposium on African vertebrates:Systematics, Phylogeny and Evolutionary Ecology (K.L. Schuchmann Ed). Mus. Alexander Koenig, Bonn, pp 479-494.

King, M. (1973) Aust. J. Zool. 21, 21-32.

King, M. (1977) Aust. J. Zool. 25, 43-57.

King, M. (1981) In: Evolution and Speciation: Essays in honour of M.J.D. White. (W.R. Atchley and D. Woodruff Ed), Cambridge University Press, pp. 262-285.

King, M. (1985) In: Biology of Australasian Frogs and Reptiles. (G. Grigg, R. Shine and H. Ehmann Eds). Surrey, Beatty and Sons, Sydney, pp. 165-175.

King, M. (1987a) Aust. J. Zool. 35, 507-531.

King, M. (1987b) Search 18, 252-254.

King, M. (1987c) Aust. J. Zool. 35, 641-654.

King, M. and D. King (1975) Aust. J. Biol. Sci. 28, 89-108.

Kluge, A.G. (1967) Aust. J. Zool. 15, 1007-1108.

Kluge, A.G. (1987) Misc. Publ. Mus. Zool. Univ. Mich. 173, 1-54.

Matthey, R.R. (1933) Rev. Suisse. Zool. 40, 281-316.

McBee, K., Sites, J.W., Engstrom, M.D. and Rivero-Blanco,
C. (1984). J. Herpetol. 18, 83-4.

McBee, K., Bickham, J.W. and Dixon, J.R. (1987). J. Herpetol. 21, 68-71.

Mengden, G.A. (1983) Chromosomal evolution and the Phylogeny of Elapid snakes. PhD. thesis, Australian National University.

Mengden, G.A. (1985a) In: The Biology and Australasian Frogs and Reptiles (G. Grigg, R. Shine, H. Ehmann Eds), Surrey, Beatty and Sons. Sydney, pp. 185-192.
Mengden, G.A. (1985b) In: The Biology of Australasian Frogs and Reptiles. (G. Grigg, R. Shine, H. Ehmann. Eds), Surrey, Beatty and Sons, pp 193-208.
Molnar, R.E. (1982) In: The Fossil Vertebrate Record of Australia. (P.V. Rich and E.M. Thompson Eds), Monash University Press, Melbourne, pp. 228-233.
Olmo, E. (1986) Animal Cytogenetics: vol. 4: Chordata 3. A. Reptilia, pp. 1-100.
Ota, H. (1988) Experientia 44, 66-68.
Peccinini-Seale, D. (1981) Genetica, 56, 123-148.
Schwaner, T.D., Baverstock, P.R. Dessauer H.C. and Mengden G.A. (1985) In: The Biology of Australasian Frogs and Reptiles. (G. Grigg, R. Shine, H. Ehmann, Eds), Surrey, Beatty and Sons, pp. 177-184.
Solleder, E. and Schmid M. (1988) Amphibia/Reptilia 9, 301-310.
Stirton, R.A., Tedford R.T. and Miller R.H. (1961) Geog. Rev. 62, 40-70.
Storr, G.M. (1964) Senck. Biol. 45, 577-589.
Tyler, M.J. (1979) In: The South American Herpetofauna: its origin, evolution, and dispersal (W.E. Duellman. Ed), Univ. Kans. Mus. Nat. His. Monog. 7: pp. 73-106.
Veevers, J.J. and M.W. McElhinny (1976). Earth-Science Rev. 12, 139-159.
Wermuth, H. (1967) Tierreich 85-127.
Witten, G.J. (1982) In: Evolution of the flora and fauna of Arid Australia (W.R. Barker and P.J.M. Greenslade, Eds), Peacock Publ. S. Australia, pp. 225-228.
Witten, G.J. (1983) Aust. J. Zool. 31, 533-540.

DNA AND CHROMOSOME EVOLUTION IN LACERTID LIZARDS

E.Olmo, G.Odierna, T.Capriglione and A.Cardone

Dipartimento di Biologia Evolutiva e Comparata, via Mezzocannone 8, I 80134 Napoli, Italy.

SUMMARY: Recently, several studies have been conducted on the karyology of lacertids by means of chromosome banding techniques and DNA analysis using restriction enzymes. The results on the localization of the nucleolar organizer and the presence and evolution of highly repetitive DNA fractions appear particularly useful for systematic research. Other methods, like the C-banding, provide interesting information on chromosome organization and composition, the mechanisms responsible for chromosome rearrangements and sex chromosome differentiation. These investigations have shown the importance of considering different types of karyological results, both the morphological and the molecular ones, and to interpret them in a wider context including also other types of data.

INTRODUCTION

During the last years, the systematics of lacertids has been subjected to a profound re-examination based at first on morphological studies (Arnold, 1973; Bohme, 1971), and subsequently on electrophoretical and immunological studies (Guillaume and Lanza, 1982; Mayer and Tiedemann, 1982; Lutz and Mayer, 1985; Busack and Maxson, 1987; Mayer and Lutz, 1989).

These investigations have concerned mainly westpalearctic genera belonging to the so-called Lacerta s.l.-complex, for which Arnold (1973) has suggested a new classification. More recently, they have been extended also to lacertids from other geographical areas (Scerbak, 1975; Bohme, 1981; Mayer and Berger-Dell´Mour, 1988).

Until few years ago, the contribution given by karyology to the systematics of lacertids was poor, since the investigations conducted with conventional cytological methods showed only slight differences among the karyotypes of the various species studied (Olmo, 1986). In recent times, the development of karyological banding techniques and of DNA analysis using restriction enzymes has provided new powerful tools of investigation. Some recent studies (Odierna et al., 1987; Olmo et al., 1986; Capriglione et al., 1989; Olmo et al., 1987a) seem to demonstrate that these techniques can be employed successfully in order to get a deeper insight into the cytogenetics of lacertids.

This paper aims at reviewing the most recent and modern results on the karyology of lacertids, in particular those obtained in our laboratory, and to assess their contribution to a better understanding of the systematics and phylogeny of lacertid lizards.

KARYOTYPE MORPHOLOGY

The peculiarity of the karyotype and its uniformity at the intrafamilial level are the two features of lacertid karyology already evidenced by conventional methods.

Compared with most saurians, where the most wide-spread

karyotype has 12 biarmed macrochromosomes and 24 microchromosomes, lacertids show a fairly peculiar karyotype constituted by 36 uniarmed macrochromosomes and 2 microchromosomes (Clmo, 1986). A similar chromosome formula is found in few other taxa, such as gekkonids and Ameiva-group teiids (Olmo, 1986). Gekkonids and lacertids being distant families, the resemblance between their karyotypes can be considered as the result of a mere evolutionary convergency. The resemblance between teiids and lacertids is of greater interest, since these two families are supposed to have a common origin (Estes, 1983)

As far as the intrafamilial uniformity is concerned, only 10 out of the 78 species so far studied show karyotypes a little different from the 38-chromosome one, as can be seen from Table 1. Some of them differ in the presence (Gallotia, Takydromus sexlineatus) or in the absence (Lacerta horvathi) of a pair of microchromosomes. Some, like L.lepida and L.parva, show Robertsonian variants characterized by one or more pairs of biarmed macrochromosomes (Clmo, 1986).

NOR-LOCALIZATION

The study of the localization of the nucleolar organizer (NOR) is considered of some diagnostic relevance and has been used many times in order toinquire into the taxonomy of different groups of vertebrates (Hsu et al., 1975; Galetti et al., 1984; Bickam and Rogers, 1985). However, cases of marked

Table 1. Diploid number (2n), fundamental number (FN), sex-chromosome morphology (Sex) and NOR localization in lacertid lizards. W-chromosome morphology: m=microchromosome, h=homo morphic heterochromatic, e=homomorphic euchromatic, X=biarmed macrochromosome, A=uniarmed macrochromosome. For the different kind of NOR localization see text. The data are from: Cano et al. (1984), Capula et al. (1989), De Smet (1981), Gilboa (1974), Gorman (1973), Kupriyanova (1989), Lopez-Jurado et al. (1986), Odierna et al. (1987), Odierna et al. (1989a, b),Olmo (1986), Olmo et al. (1987a, b), Palacios and Elvira (1979), Peccinini-Seale (1981), Rykena and Nettmann (1986), this paper.

SPECIES	2n	FN	Sex	NOR
LACERTA part I:				
L.agilis	38	38	m	L
L.lepida	36	38	h/m	L
L.pater	36	38		L + MS
L.princeps	36	38		L
L.schreiberi	38	38		L
L.strigata	38	38	m	
L.trilineata	38	38	m	L
L.viridis	38	38	h/m	L
LACERTA part II:				
L.armeniaca	38	38	m	
L.bedriagae	38	38		L
L.caucasica	38	38		
L.dahli	38	38		L
L.derjugini	38	38		
L.dugesii	38	38	m	L
L.graeca	38	38		M
L.horvathi	36	36	h	L
L.laevis	38	38		
L.monticola	38	38	h	MS
L.mixta	38	38		
L.oxycephala	38	38		
L.parva	24	38		m
L.parvula	38	38		
L.perspicillata	38	38		
L.portschinskii	38	38		
L.praticola	38	38		
L.raddei	38	38		
L.rostombekovi	38	38		
L.rudis	38	38		
L.saxicola	38	38		
L.unisexualis	38	38		
L.vivipara	36	36	X/A	MS
PODARCIS:				
P.erhardii	38	38	m	
P.filfolensis	38	38		
P.hispanica	38	38		
P.lilfordi	38	38		
P.melisellensis	38	38	e/m	M
P.milensis	38	38		

P.muralis	38	38	e	M
P.peloponnesiaca	38	38		M
P.pityusensis	38	38		
P.sicula	38	38	e/m	M
P.taurica	38	38		M
P.tiliguerta	38	38	e	M
P.wagleriana	38	38	e	M
ALGYROIDES				
A.marchi	38	38		M
A.moreoticus	38	38		M
A.nigropunctatus	38	38		M
GALLOTIA:				
G.atlantica	40	40		
G.galloti	40	40	h	S
G.simonyi	40	40		S
PSAMMODROMUS				
P.algirus	38	38	h/m	M
P.hispanicus	38	38		
EREMIAS:				
E.arguta	38	38	m	
E.grammica	38	38		
E.guttulata	38	38		
E.multiocellata	38	38		
E.olivieri	38	38	m	
E.persica	38	38		
E.regeli	38	38		
E.strauchi	38	38		MS
E.velox	38	38	m	
HELIOBOLUS:				
H.lugubris	38	38	m	MS
PEDIOPLANIS:				
P.husabensis	38	38		MS + S
P.namaquensis	38	38	m	MS + S
P.undata	38	38		MS + S
MEROLES:				
M.cuneirostris	38	38	h	MS
M.suborbitalis	38	38		MS
ACANTHODACTYLUS:				
A.boskianus	38	38		
A.erythrurus	38	38	h	MS
A.opheodurus	38	38		
A.pardalis	38	38		
A.schreiberi	38	38		
A.scutellatus	38	38		
OPHISOPS:				
O.elegans	38	38	m	
TAKYDROMUS:				
T.amurensis	40	40		
T.formosanus	38	38		
T.septentrionalis	38	38		
T.sexlineatus	42	42	h	M
T.smaragdinus	38	38		
T.tachydromoides	38	38		

variability of this feature are known, with some variations, also at the intraspecific level (Nardi et al., 1978; Pillips et al., 1988), and therefore caution should be used when conducting this kind of investigation. Moreover, it cannot be ruled out that alsc in the case of the NOR, some resemblances are the result of evolutionary convergency (Amemiya and Gold, 1988).

NOR localization is known in 37 species of lacertids (Table 1); in most cases it has been detected by the AgNOR technique proposed by Howell and Black (1980). In particular, 31 species have been studied by us.

The results obtained by these studies seem to provide interesting information on lacertid cytotaxonomy, as already observed by Odierna et al. (1987). In this family, though taking into account the difficulty in distinguishing exactly between the different pairs of chromosomes, it is possible to identify, with good reliability, at least four different types of NOR-bearing chromosomes according to their different sizes: m = microchromosomes; S = small chromosomes, which correspond to a pair included between the 18th and the 16th; MS = medium small chromosomes, which correspond to a pair included between the 16th and the 14th; M = middle size chromosomes, which correspond to a pair included between the 12th and the 10th; and L = large chromosomes, which correspond to a pair included between the 7th and the 5th (Fig.1).

Most lacertids have a single pair of NOR-bearing chromosomes; only <u>L.pater</u> and the three species of <u>Pedioplanis</u> studied show two of them (Odierna et al., 1989a,b). The NOR but

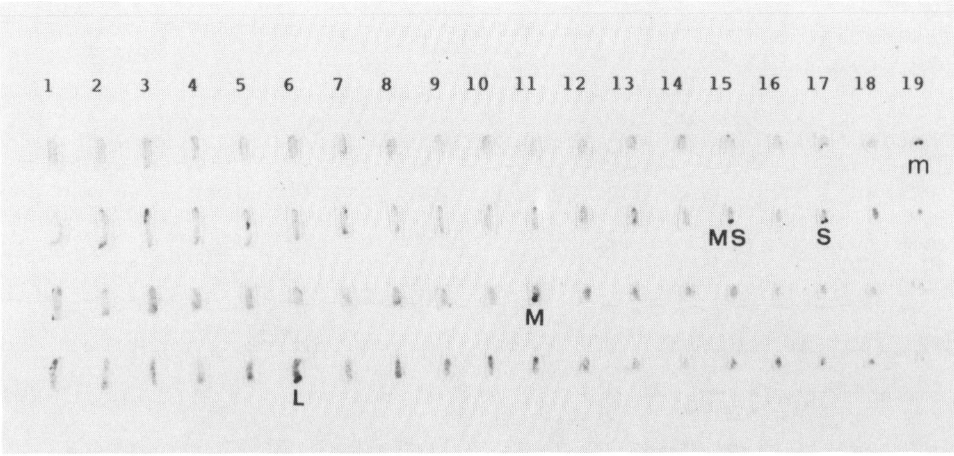

Fig. 1. Karyotypes showing the different types of NOR found in lacertid lizards. See text for explanation.

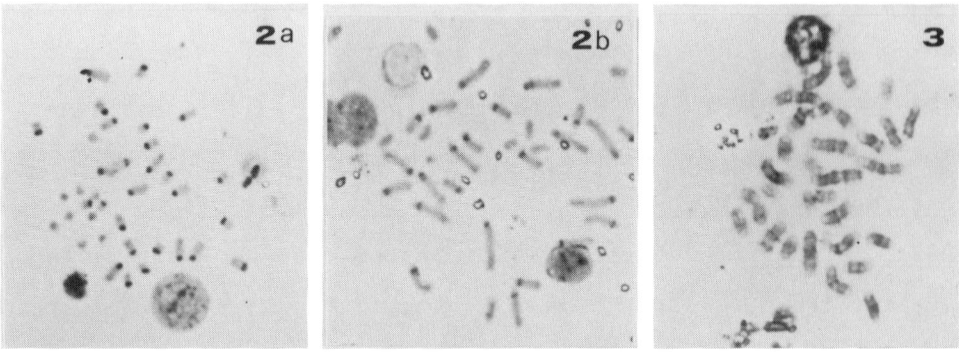

Fig. 2. Centromeric (a) and(b)paracentromeric C-banding pattern in P. sicula.
Fig. 3. G-banding pattern in P.sicula.

few exceptions is localized at the telomeric level.

The distribution of the different types of NOR-bearing chromosomes shows that the species belonging to the same clearly defined taxonomic unit usually have the NOR localized on the same pair of chromosomes. It is the case of *Podarcis*, *Gallotia* and *Pedioplanis*. These observations are often consistent with the results obtained with morphological electrophoretical and immunological investigations, and suggest that the study of NOR localization is a useful and reliable method of detecting the relationships between species ascribed to the same genus, or however quite closely related.

The interpretation of the data appears more difficult when a comparison is made between species which are supposed to belong to distant or however less closely related genera. In some cases, there is a satisfactory correspondence between resemblance in NOR localization and morphologic enzymatic or immunologic affinity. This is the case of *L.graeca* and *Podarcis*, which have the NOR on the same M pair and appear very close from the immunological standpoint (Mayer and Lutz, 1989). Likewise, most of the species from the so-called Eremias-complex, such as *Estrauchi*, *H.lugubris*, *Pedioplanis*, *A.erythrurus* and *M.suborbitalis*, have the NOR (or in the case of *Pedioplanis* at least either NOR) on the same MS pair. However, a similarity in NOR localization is observed in species clearly distant from the taxonomic standpoint, such as *Podarcis* and *Psammodromus*; and, on the other hand, cases of different NOR localization are found in species like *L.dugesii* and *Podarcis*, that appear to be related from the immunological standpoint (Mayer and Lutz, 1989). This suggests that, also in lacertids, a similarity in NOR localization does not always depend on an actual phyletic affinity, but it may be due to a process of evolutionary convergency, as happens for many other morphological and karyological features.

C-BANDING

Lacertid karyology has been studied also by the C-banding method, which evidences constitutive heterochromatin (Sumner, 1972).

Several species have been investigated by this method, but the results do not seem to be useful for a comparative analysis, since, in lacertids, C-banded heterochromatin shows a wide interspecific variability and, in the species studied in greater detail, like *P.sicula*, also intraspecific variability (Olmo et al.,1985).

These studies would seem to provide more interesting information on chromosome organization and composition, some mechanisms responsible for chromosome rearrangements, and, in particular sex chromosome differentiation.

In lacertids, constitutive heterochromatin ranges from a minimum of only 4% of the total genome in *Gallotia galloti* to a maximum of more than 20% in *Podarcis sicula*, and appears mostly as centromeric or pericentromeric bands (Olmo et al., 1988). The telomeric bands are rarer and usually less intense, whereas intercalary bands are very seldom found (Clmo et al., 1985). The two patterns with centromeric and pericentromeric bands are the most common (Fig.2a,b). They are found in related species, such as L.viridis and L.trilineata, and also in different populations from the same species, as in P.sicula (Olmo et al., 1985). The former pattern is found essentially in species richer in heterochromatin.

All the C-bands intensely stain with DAPI (Olmo et al., 1988), a fluorochrome specific for chromosome areas containing DNA rich in adenine-thymine (Schweizer, 1980).

Like in many other cases, also in lacertids, the C-bands are areas where highly repetitive DNA fractions are preferentially localized. In fact, in Podarcis a satellite DNA has been observed to be localized on the centromeric bands (Capriglione et al., 1989) and another one is probably on the telomeric bands (Cardone et al., 1989).

C-banding evidences a peculiar aspect of sex chromosomes. In fact, some species (see Table 1) show sex chromosomes where the two homologues have the same shape and size, but W is different in being intensely and completely heterochromatic (Olmo et al., 1987a). The situation of sex chromosomes is heterogeneous. In fact, in other species sex chromosomes are either homomorphic and euchromatic, or heteromorphic W being a microchromosome.

OTHER TYPES OF BANDING PATTERNS

Studies using other banding methods are less advanced and no results have been published so far.

In our laboratory, several methods are being used, such as G-banding, replication banding and the restriction enzyme analysis. So far the results are preliminary and do not allow a comparative analysis; however, they show that it is possible to employ any banding method in the study of lizard karyology with results that can be compared with those obtained so far in mammals. Fig.3 shows the results that can be obtained with these methods.

DNA ANALYSIS

In lacertids, uniformity in karyotype morphology is accompanied by uniformity in genome sizes, which are usually quite small, ranging between 2.2 pg/N in *L.vivipara* and about 6 pg/N in *Psammodromus algirus* . However, it isto be noted that the data in literature are conflicting, and, if we consider the results obtained by a single author, the range appears even more reduced (Olmo, 1986).

The genome composition of lacertids appears also quite uniform. Almost all the species studied show the same percentage of guanine-citosine and a very similar amount of single-copy DNA (which is known to include most of the structural genes) (Olmo,

1981; Capriglione et al., 1986). Greater differences between the various species are observed in repetitive DNA fractions (Olmo et al., 1988).

A detailed analysis of the genome of P.sicula with restriction enzymes (Capriglione et al., 1989) has evidenced some homogeneity also at the intragenomic level with prevalence of very similar restriction sites and of sequences quite rich in adenine-thymine. Another relevant feature is the presence of two highly repetitive satellite DNAs which can be evidenced with Taq 1 (Capriglione et al., 1989) and Hind III (Cardone et al., 1989) respectively. Both sequences are localized on constitutive heterochromatin; however, Taq 1 is found at the level of the centromeric C bands (Capriglione et al., 1989), while the other sequence seems to be present at the telomeric level (Cardone et al., 1989). The Taq 1 sequence is more wide-spread and is found in all the species from the genus Podarcis, as well as in L.dugesii, even though in the latter species it would have undergone a more marked divergence in composition (Fig.4). The distribution of the Hind III sequence is not yet clearly determined.

KARYOTYPE EVOLUTION: TREND AND MECHANISMS

The origin of the karyotype of lacertid lizards is still a debated question. According to some authors, the ancestral karyotype of all saurians would have possessed all uniarmed

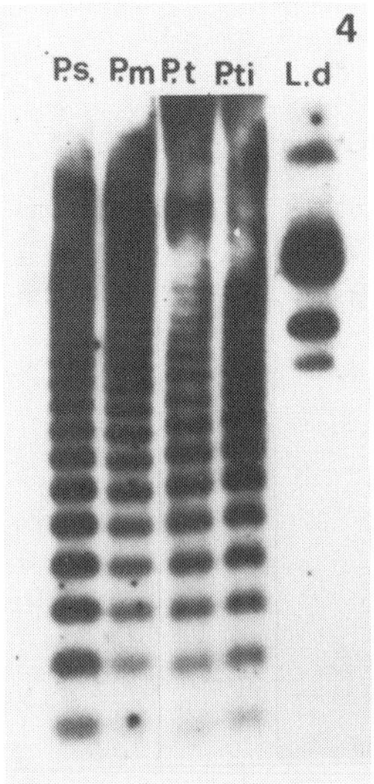

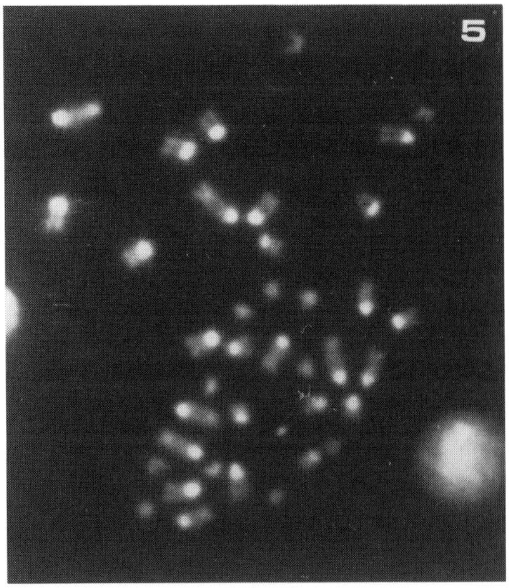

Fig.4. Hybridization of Taq 1 DNA from P.sicula, with the DNA of other species: P.s.= P.sicula, P.m.= P.muralis, P.t.= P.taurica, P.ti= P.tiliguerta, L.d.= L.dugesii.
Fig.5. DAPI stained chromosomes from P.sicula.

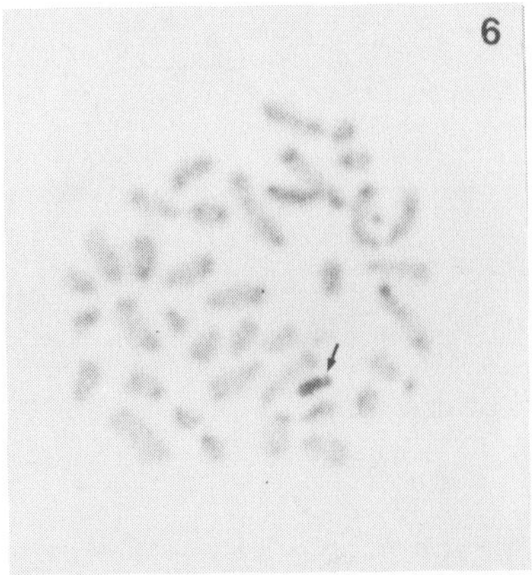

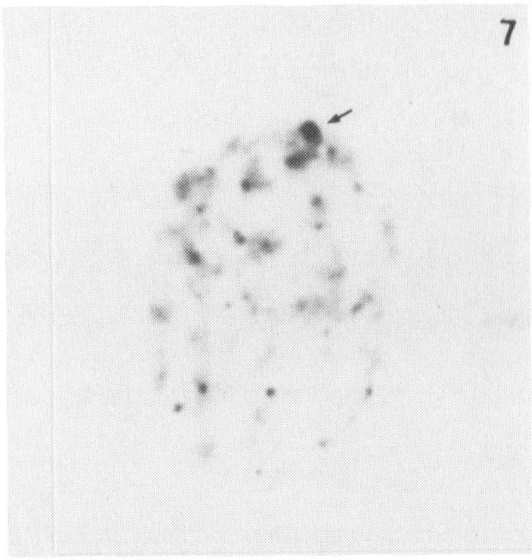

Fig.6. Homomorphic heterochromatic W sex-chromosome in G.galloti.
Fig.7. Meiosis of L.viridis. Note the sex-bivalent with the heterochromatic W.

elements, like that of gekkonids and lacertids (King, 1981). However, this does not agree with the presence, in almost all the other species, of equal karyotypes with 12 metacentric chromosomes and 24 microchromosomes, which, on the above hypothesis, would have originated independently as a consequence of evolutionary convergency.

Another hypothesis, which at present better agrees with the available data, is that the ancestral karyotype of all saurians was the one with 12 biarmed chromosomes and 24 microchromosomes (Paull et al., 1976). The karyotype of lacertids would be therefore derived. Gorman (1970) has stressed the close resemblance between the karyotype of lacertids and that of _Ameiva_-group teiids (a family related to lacertids, Estes 1983), which, in turn, would have originated from the 12+24 one of _Dracaena_-group teiids. According to this author, the karyotype of lacertids would have originated from that of _Ameiva_-group teiids with few modifications.

Independently of its origin, the lacertid karyotype with all acrocentric elements seems to be more wide-spread than the ones possessing one or more pairs of biarmed macrochromosomes. Moreover, some observations seem to indicate that the evolution of lacertid karyotype involved a small reduction in the number of microchromosomes, mainly through processes of translocation of these elements onto the macrochromosomes (Olmo et al., 1987b). In fact, species belonging to different lineages and considered ancient, such as _Takydromus_, _Gallotia_ and _Psammodromus_, show higher diploid numbers and/or either stable or supernumerary microchromosomes (Olmo et al., 1985; Olmo et al., 1987). In addition, in many of these species the NOR is localized on a pair of microchromosomes, whereas in more advanced species with a lower diploid number, the NOR is always localized on the macrochromosomes (Odierna et al., 1987).

The formation of biarmed macrochromosomes through a process of Robertsonian fusion is an extremely rare event, which so far has been observed only in L.parva, L.lepida and its allied, and in the sex chromosomes of L.vivipara.

Therefore, lacertids appear as a group with a very conservative karyotype which, during millions of years, has undergone almost only limited phenomena of loss of some microchromosomes or of their translocation onto the macrochromosomes.

According to some hypotheses, like the canalization theory (Bickham and Baker, 1979), this karyological uniformity would indicate that lacertid lizards have evolved the optimum karyotype for their adaptive zone (i.e., the karyotype which better contributes to the fitness of the individuals).

The conservativeness of lacertid karyotype, however, might be due to peculiar features of their chromosome composition. Several authors have quite frequently observed that metacentric chromosomes show pericentromeric C-bands that stain with chromomicine A3 and hence contain some CG-rich DNA; they have also noticed that, on the contrary, acrocentric chromosomes show centromeric C-bands that stain with DAPI and hence contain some AT-rich DNA (Mayr et al., 1984, John and King, 1977; John et al., 1985; King et al., 1986). Mayr et al. (1984) suggest that this is not a mere coincidence, but there can be a precise correlation between centromeric heterochromatin and chromosome morphology.

In this regard, it is noteworthy that, in almost all the lacertids studied, acrocentric chromosomes show centromeric or pericentromeric C-bands that intensely stain with DAPI (Fig.5) (Olmo et al., 1988); moreover, these bands contain highly repetitive DNA with a sequence showing long stretches of adenine-thymine (Capriglione et al., 1989). On the contrary, heterochromatin is extremely scarce in biarmed chromosomes of L.lepida (Olmo et al., 1986).

Variation in the amount of constitutive heterochromatin, the meaning of which is stillunknown, is another chromosome mutation observed in lacertids; it is probably accompanied by the amplification of some highly repetitive DNA sequences (Olmo et al., 1986; Capriglione et al., 1989). In Podarcis s.sicula,intraspecific variations are observed in the amount and localization of the C-bands, which show a trend that agree quite well with the distribution of the various populations of this species. It has been hypothesized that such variations may be related to speciation phenomena (Olmo et al., 1986; Olmo, 1987).

As has already been mentioned, in lacertids, the C-bands and probably most of the highly repetitive sequences are localized almost exclusively at the centromeric and telomeric levels (Capriglione et al., 1989; Cardone et al., 1989).

Moreover, it has been observed that centromeric heterochromatin contains AT-rich DNAs, the presence of which seems to be related to the uniarmed condition of the chromosomes, and that microchromosomes seem to translocate preferentially onto the telomeres of the macrochromosomes (Odierna et al., 1987). All these observations seem to indicate that, in these organisms, the centromere and the telomere are the most, if not the only, regions involved in chromosome rearrangements; instead the other regions of the chromosome, being perhaps richer in gene sequences, would seem more conservative. This hypothesis, however, needs to be verified by the study of chromosome composition with methods allowing a more detailed examination of the intercalary areas, such as the G-banding, the replication bands and the restriction enzyme analysis.

SEX CHROMOSOMES

In lacertids, heterochromatin seems to have played a paramount role in the differentiation of sex chromosomes. In this family (see Table 1), as in other reptiles, besides species with either undistinguishable or clearly heteromorphic sex chromosomes, there are species showing female heterogamety, where the two sex chromosomes have the same shape and size, but W is completely heterochromatic and C-band positive (Olmo et al., 1987a) (Fig.6). This type of heterogamety is considered primitive in several reptiles, and probably so it is also in lacertids compared with another type of sex chromosomes found in other species, and sometimes also in other populations from the same species, where W is smaller than Z or is even a microchromosome. Two interesting features of the heterochromatic W are its intense stainability with DAPI, which indicates that it contains much AT-rich DNA (Olmo et al., 1987a), and its behaviour at meiosis, when it pairs with Z only along a short euchromatic terminal stretch and the rest remains heterochromatic, strongly spiralized and unpaired (see Fig.7, Olmo et al., unp.). These observations are consistent with the hypothesisadvanced by Singh et al. (1980) at first for snakes, but afterwards also for other organisms, that sex chromosome differentiation would start with the accumulation of AT-rich highly repetitive DNA on either homologues. Perhaps by bringing about heterochrony in the duplication and in the spiralization cycle between the two homologues, this DNA would prevent crossing-over in the area carrying the sex determinant, and hence would contribute to its preservation (Jones, 1989).

In lacertids, in most cases the subsequent stage of sex chromosome differentiation lies in a progressive deletion of W which becomes smaller and smaller to turn into a microchromosome.

The presence of different types of sex chromosomes in several populations from the same species, like in L.lepida and L.viridis, indicates that, in this family, sex chromosome differentiation has taken place repeatedly and independently in different taxa (Olmo et al., 1987a). Interestingly, L.vivipara shows a heterogamety of the Z1Z2W type where W is a biarmed chromosome derived by centric fusion (Chevalier, 1969). This would be indicative of the existence in lacertids of different mechanisms of sex chromosome differentiation.

KARYOLOGY AND SYSTEMATICS

The data on lacertid karyology so far obtained, though not allowing a complete outline of the karyological evolution of the family, however provide interesting information on the taxonomy of several genera, as well as on the phyletic relationships between the various species, mainly the ones ascribed to the Lacerta s.l.-complex.

Taking into account mainly the karyological features that can be considered apomorphic, it is possible to confirm or further support the reliability and the uniformity of some genera, like *Gallotia*, which includes all species with 38 macrochromosomes and 2 microchromosomes and the NOR localized on a pair of S; *Pedioplanis* including all species with the two NORs localized one on a MS and the other on a S chromosome; and *Podarcis* where all the species have certain highly repetitive DNAs and the NOR on a pair of M chromosomes.

It is also possible to give some support to the reliability of some species complexes, such as the so-called Lacerta s.l.-complex and the Eremias-like lacertids, as well as to advance a hypothesis on some stages of karyological evolution in lacertids.

A first distinction on a karyological basis can be made among the genus *Takydromus* that shows a chromosome formula with 2n=40 or 42 and a greater number of microchromosomes; the genus *Gallotia* that has a formula with 2n=40 and 38 macrochromosomes; and all the other genera of the family possessing no more than 38 chromosomes with 36 macrochromosomes and only two microchromosomes. This is constistent with the observations based on morphological immunological and electrophoretical data, which suggest that *Takydromus* is a sister-group of the other lacertids (Arnold, 1984) and *Gallotia* would have diverged from the other groups of the family very early (Arnold, 1973; Lopez-Jurado et al., 1980; Lutz and Mayer, 1985).

As regards the species with 38 chromosomes, there is no karyological indication on the position of *Psammodromus*, which is considered quite an ancient group. On the contrary, the distinction between the Eremias-like lacertid including all species with the NOR (or at least one of them)on a MS chromosome, and the so-called Lacerta-complex, where most species have the NOR on a L chromosome, is quite clear.

In the former group the most advanced situation is found in *Pedioplanis*, which possesses a second NOR on a S chromosome.

The position of *L.vivipara*, *L.monticola* and *L.parva* within the Lacerta s.l.-complex cannot be clearly defined. They show peculiar karyological features quite distinct from those of the other species. *L.lepida* and its allied (*L.princeps* and *L.pater*), for the presence of a pair of biarmed chromosomes, and a group including Podarcis, L.graeca, Algyroides and L.dugesii, about which there are several data on karyotype morphology and DNA composition, can be also considered as two distinct groups. In the latter group, the most primitive species is supposed to be L.dugesii, which still has a NOR-bearing L chromosome, but has a satellite DNA present also in Podarcis and its allied and not in the other species of Lacerta. From this lineage, the origin of Podarcis, Algyroides and L.graeca seems to have involved also the NOR translocation from a L to a M chromosome. These observations are not yet complete and definite enough, however they agree well with many of the results obtained with morphological electrophoretical and immulogical investigations (Bohme, 1971; Arnold, 1973; Lutz and Mayer, 1985; Mayer and Lutz, 1983).

It is important to emphasize that one should not only take into account thedata on karyotype morphology, but also other karyological data, such as DNA organization and composition. An interesting example is provided by L.dugesii. In fact, on the basis of NOR localization this species would seem more closely related to species like those belonging to Lacerta s.str. or to L.lepida and its allied. Vice versa, DNA analysis clearly evidences its relationship with Podarcis, and this better agrees with the morphological and immunological data (Richter, 1979; Mayer and Lutz, 1989).

CONCLUSIONS

In these last years, more and more information has been enhancing the knowledge of lacertid karyology, essentially thanks to the employment of the most up-to-date cytological and molecular techniques.

The data obtained with these new techniques, though sometimes appearing preliminary, are nevertheless promising, and show the possibility of improving our knowledge of the trend of lacertid evolution, of the role that chromosome and genome variation would have played in this process, as well as of the mechanisms responsible for the various chromosome rearrangements, mainly the ones concerning sex chromosome differentiation.

In this regard, however, it is important to consider different types of karyological data, both the morphological and the molecular ones, and to interpret them in a wider context including also other types of data, such as the morphological electrophoretical and immunological ones.

REFERENCES

Amemiya, C.T., and Gold, J.R. (1988) Genetica. 76, 81-90.
Arnold, E.N. (1973) Bull. Br. Mus. Nat. Hist. (Zool.) London. 25, 291-366.
Arnold, E.N. (1984) In: The structure, development and evolution cf reptiles (M.W.J. Ferguson Ed.) Symp.zool.Soc. London n. 52. Academic Press, London, Orlando, pp. 47-85.
Bickham, J.W., and Baker, R. (1979) Bull. Carnegie Mus.Nat.Hist. 13,70-84.
Bickham, J.W., and Rogers, D.S. (1985) Genetica, 67, 171-184.
Bohme, W. (1971) Z.zool.Syst.Evolut.forsch. 9, 187-223.
Bohme, W. (1981) In: Handbuch d. Reptilien und Amphibien Europas Bd.1: Echsen. (Bohme W. Ed.) Akad. Verlages. Wiesbaden, pp. 479-491.
Busack, S.D., and Maxson, L.R. (1987). Amphibia-Reptilia. 8, 383-392.
Canc, J., Baez, M., Lopez-Jurado, L.F., and Ortega, G. (1984) J.Herpetol. 18, 344-346.
Capriglione,T., Olmo, E., Odierna, G., Improta, B. and Morescalchi, A. (1987) Bas.Appl.Histochem. 31, 119-126.
Capriglione, T., Olmo, E., Odierna, G., Smith, D.I. and Miller, O.J. (1989) Genetica (In press).
Capula, M., Lapini, L., and Capanna, E. (1989) Genetica 79, 11-16.
Cardone, A., Capriglione, T., Odierna, G., Redi, C., and Garagna, S. (1989) This volume.
Chevalier, M. (1969) Comp.Rend.Acad.Sci. D268, 2098-2100.
De Smet, W.H.O. (1981) Acta zool. pathol. Antverpiensia. 76, 73-118.
Estes, R. (1983) In: Advances in Herpetology and Evolutionary Biology (Rhodin, A.J. and Miyata, K. Eds.) Mus.Comp.Zool., Harvard Univ. Cambridge, Mass. pp. 365-398.
Galetti, P.M. Jr., Foresti, F., Bertollo, L.A.C. and Moreira, O. Fo. (1984) Caryologia 37, 401-406.
Gilboa, I. (1974) In: Yearbook of herpetology (Dowling, H.G. Ed.). Hiss Publ. Amer.Mus.Nat.Hist., Mew York. pp. 91-156.
Gorman, G.C. (1970) Copeia. 1970, 230-245.
Gorman, G.C. (1973) In: Cytotaxonomy and Vertebrate Evolution (Chiarelli, A.B. and Capanna, E. Eds.). Academic Press, London, New York. pp. 349-424.
Guillaume, C.P. and Lanza, B. (1982) Amphibia-Reptilia. 4, 361-375.
Howell, W.M. and Black, D.A. (1980) Experientia. 36, 1014.
Hsu, T.C., Spirito, S.E., and Pardue, M.L. (1975) Chromosoma. 53, 25-36.
John, B., and King, M. (1977) Chromosoma. 65, 59-79.
John, B., King, M., Schweizer, D., and Mendelak, N. (1984) Chromosoma. 91, 185-200.
Jones, K. (1989) In: Evolutionary mechanisms in sex determination. (Wachtel, S.S. Ed.).
King, M. (1981) In: Evolution and Speciation (Atchley, W.R., and Woodruff, D.S.) Cambridge Univ. Press, Cambridge, London, New York, New Rochelle, Melbourne, Sydney. pp. 262-285.King, M.,

King, M., Honeycutt, R., and Contreras, N. (1986) Genetica. 70, 191-201.
Kupriyanova, L. (1989) This volume.
Lopez-Jurado, L.F., Cano, J., and Baez, M. (1986) Amphibia-Reptilia. 7, 259-270.
Lutz, D., and Mayer, W. (1985) Amphibia-Reptilia. 7, 53-61.
Mayer, W., and Berger-Dell'Mour, H. (1988) Herpetozoa. 1, 23-29.
Mayer, W., and Lutz, D. (1989) Z.f.zool.Syst.Evolut.forsch. (In press).
Mayer, W., and Tiedemann, F. (1982) Amphibia-Reptilia. 2, 349-356.
Mayr, B., Schweizer, D., and Geber, G. (1984) Jour. Hered. 75, 79-80.
Nardi, I., Barsacchi-Pilone, G., Batistoni, R., and Andronico, F. (1978) Chromosoma. 64, 67-84.
Odierna, G., Capriglione, T., and Olmo, E. (1989 a) In: Evolution of terrestrial vertebrates. (Ghiara, G.F., Angelini,F., and Olmo, E. Eds.) Mucchi, Modena (In press).
Odierna. G., Olmo, E., Capriglione, T., and Caputo, V. (1989 b) J. Herpetol. (In press).
Odierna, G., Olmo, E., and Cobror, C. (1987) Amphibia-Reptilia. 8, 373-382.
Olmo, E. (1981) Genetica. 57, 39-50.
Olmo, E. (1986) In: Animal Cytogenetics. 4 Chordata 3A. (John, B. Ed.) Gebruder, Borntraeger, Berlin, Stuttgart.
Olmo, E. (1987) Boll.zool. 54, 49-54.
Olmo, E., Odierna, G., and Capriglione, T. (1987 a) Chromosoma. 96, 33-38.
Olmo, E., Odierna, G., and Capriglione, T. (1987 b) In: Proc. 4th Ord. Gen. Meeting Soc.Eur.Herp. (van Gelder, J.J., Strijbosch, H., and Bergers, P.J.M. Eds.). Nijmegen 1987. Faculty of Sciences. Nijmegen. pp. 303-306
Olmo, E., Odierna, G., Capriglione, T., and Merciai, B.M. (1988) Comp. Biochem. Physiol. 89b, 1-4.
Olmo, E., Odierna, G., and Cobror, O. (1986) Genetica. 71, 63-74.
Palacios, F., and Elvira, B. (1979) Donana Acta Vertebr. 6, 217-220.
Peccinini-Seale, D. (1981) Genetica. 56, 123-148.
Paull, D., Williams, E.E., and Hall, W.P. (1976) Breviora. 441, 1-31.
Phillips, R.B., Pleyte, K.A., and Hartley, S.E. (1988) Cytogenet. Cell Genet. 48, 9-12.
Richter, K. (1979) Zool.Abh. Dresden. 36, 1-9.
Rykena, S., and Nettmann, H.K. (1986) In: Studies in Herpetology (Rocek, Z. Ed.) Charles Univ., Prague. pp. 193-196.
Scerbak, N.N. (1975) Katalog afrikaniskih jascurok; Kiew (Akad. Nauka)
Schweizer, D. (1980) Cytogenet. Cell Genet. 27, 190-193.
Singh, L., Purdom, I.F., and Jones, K.W. (1980) Chromosoma 79, 137-157.
Sumner, A.T. (1972) Expl. Cell Res. 75, 304-306.

Cytogenetics of Amphibians and Reptiles
Advances in Life Sciences

PATTERNS AND PROCESSES OF SEX CHROMOSOME EVOLUTION IN GEKKONID LIZARDS (SAURIA: REPTILIA).

Craig Moritz

Department of Zoology, University of Queensland, St. Lucia QLD 4067 Australia

SUMMARY: Gekkonid lizards vary in their mode of sex determination and for the presence of heteromorphic sex chromosomes (HSCs). The sporadic distribution and limited differentiation of HSCs suggests that they are of recent origin. Detailed analyses of two genera with highly variable HSCs, *Gehyra* and *Heteronotia*, suggest that structural rearrangement of the heteromorphic element precedes both the invasion of euchromatin by heterochromatin and changes in the timing of condensation. In contrast to the situation in snakes and some other groups of lizards, this observation reinforces that notion that the fundamental step in sex chromosome evolution is to isolate a segment from recombination, and emphasizes that there are several cytological mechanisms that can initiate such isolation. The variation for sex determining mechanisms and for the presence and form of HSCs makes this an ideal group for investigating the evolution and molecular basis of sex determination.

The differentiation of sex chromosomes is regarded as a significant, and possibly irreversible, step in the evolution of sex determination (Bull, 1983; Bull and Charnov, 1985). The early stages in the differentiation of heteromorphic sex chromosomes (HSC's) remain poorly documented despite the fact that HSCs are widespread in animals and have arisen on many separate occasions (Ohno, 1967; White, 1973). Comparative studies of sex chromosomes, among snakes in particular (Becak and Becak, 1969; Ray Chaudhuri et al., 1970; Singh et al., 1976, 1980; Mengden, 1981; Becak, 1983), have suggested two models for the primary stages of sex chromosome differentiation. One model suggests that the initial step is to isolate from recombination a segment containing one or more major sex-determining genes (Darlington, 1958; Ohno, 1967). Within such a segment, the accumulation of other genes associated with sex-determination may be favoured and there may be a tendancy to accumulate slightly deleterious mutations, or to amplify repetitive sequences (Charlesworth, 1978; Bull, 1983). Alternatively, Jones (1983, 1984; Jones and Singh, 1985) suggested that changes in the timing of condensation or replication precede structural rearrangement and the suppression of recombination.

These models, and others concerned with changes in the mode of sex determination (see Bull, 1983), can be evaluated by comparative studies of groups with labile systems of sex-determination and variable sex chromosomes. The gekkonid lizards are promising material for such studies because both environmental (ESD) and genetic (GSD) sex determination have been reported within the family; because HSC's in early stages of differentiation are evident in some groups; and because sex chromosomes have differentiated several times in the evolutionary history of this family (King, 1977; and see below).

This paper briefly reviews the known cases of ESD and HSC's in geckos. Particular attention is given to cases where HSC's are in the very early stages of differentiation, i.e., where there is variation within or between populations for the presence or form of HSC's and where the primary stages of structural differentiation can be identified.

THE DISTRIBUTION OF ESD AND HSC's

Environmental sex determination has only been reported in two species, the eublepharine *Eublepharis macularius* (Wagner, 1980; Bull, 1987) and the gekkonine *Gekko japonicus* (Tokunaga, 1985). In the former only males are produced at incubation temperatures above 31 C and only females are produced below 26 C. In contrast, males of *Gekko japonicus* predominate at intermediate temperatures (28 C) and are rare at low (24 C) and high (32 C) temperatures. This report is of particular interest because of an earlier observation of well differentiated sex chromosomes in this species (see below). There is an obvious need to determine whether temperature effects operate on populations of this gecko with sex chromosomes. There have been no reports of the absence of temperature effects on sex determination in geckos; such observations would help to identify constraints on the evolution of ESD (Bull, 1983).

Cytologically distinguishable sex chromosomes (i.e., HSC's) have been observed in several genera of gekkonine and one species of sphaerodactyline gecko (Table I). There have been no reports of sex chromosomes in the Diplodactylinae, where males and females have been examined for 11 of the

TABLE I

Reports of heteromorphic sex chromosomes in geckos. N_f and N_m refer to the numbers of females and males examined, respectively. The plus symbols indicate that the HSCs differ in the amount of heterochromatin (Het.) or by structural rearrangements (S.R.)

Species	N_f/N_m	Type	Het.	S.R.	Reference
GEKKONINAE					
Phyllodactylus lanei	2/1	ZW	?	+	King, 1981
P. marmoratus	15/18	ZW	?	+	King and Rofe, 1976
Gehyra australis	6/5	ZW	+	-	King, 1977, 1983
G. nana*	49/38	ZW	-	-	Moritz, 1986
G. purpurascens	30/42	ZW	-	+	Moritz, 1984a
Heteronotia binoei	50/38	ZW	-	+	Moritz, 1984b; this paper
H. sp.	12/9	ZW	-	+	Moritz, et al., 1990
Gekko gecko	6/4	XY	-	?	Solleder and Schmid, 1984
G. japonicus	1/1	XY	+	+	Yoshida and Msahiro, 1974
SPHAERODACTYLINAE					
Gonotodes ceciliae	0/2	XY	?	+	McBee et al., 1987

* The W chromosome in Gehyra nana is marked by an additional NOR site (see text)

92 species (King, 1987). Both male and female heterogamety have been reported for geckos and each mode is represented within the Gekkoninae (Table I). The distribution of HSC's appears to be sporadic; there are no genera where all species have HSC's and there are no examples of sister-species sharing HSC's. This pattern attests to the recent differentiation of sex chromosomes among geckos, although these data should be viewed with caution as few species have been studied with banding techniques. A recent report of HSC's in Lepidodactylus lugubris is excluded from Table I. Volobouev and Pasteur (1988) found that one chromosome pair was consistently

heterozygous for BrdU-replication-banding pattern in females. However, because this species is parthenogenetic (all-female) and presumably has clonal inheritance, fixed chromosomal heterozygosity in females does not indicate the presence of HSC's; it may simply reflect fixed clonal heterozygosity.

In some of the cases listed in Table I, e.g., P. lanei, G. japonicus, and *G. cecilidae*, additional sampling is needed to confirm that the observed heterozygosity does represent HSCs. However, only small samples are required. With the null hypothesis being autosomal inheritance, the probability that all n randomly sampled individuals of the "heterozygous" sex will be heterozygous is $(2pq)^n$, which for an estimated allele frequency of 0.5 (among individuals of that sex) is 0.5^n. Thus, a sample of five members of one sex in which all individuals are heterozygous is sufficient to reject ($p < 0.01$) autosomal inheritance in its simplest form. One explanation for such a deviation is the presence of HSC's; others include clonal inheritance or lethality of homozygotes, (e.g., MacGregor and Horner, 1980).

One significant exception to the sparse distribution of HSC's is the flap-footed lizards of the family Pygopodidae. It has recently been suggested that these are most closely related to either diplodactyline (Kluge, 1987) or carphylodactyline (King, this volume) geckos, so it is appropriate in the present context to consider them along with previously recognized members of the gekkonidae. King (1977, and this volume) has found that HSC's are widespread among pygopodids and has identified some highly derived HSC's and multiple sex chromosome systems in males.

Consistent with the phylogenetic evidence for their recent evolution, the HSC's of geckos (other than pygopodids) have limited differentiation. Indeed, it is usually possible to identify the primary steps in differentiation by comparing heteromorphic elements with each other and with the homomorphic progenitors. Of the HSC's reported, only one appears to have heterochromatin amplification as the initial step and this involves short-arm addition rather than invasion of euchromatin (King, 1977). The most strongly differentiated HSCs, those reported for G. japonicus, appear to differ in both heterochromatin content and size (Yoshida and Msahiro, 1974). The remainder involve some form of structural rearrangement as the initial step (Table I), although there is no C-banding reported for Gonotodes ceciliae or for Phyllodactylus lanei. For P. marmoratus the sex chromosomes were compared by G-banding, not C-banding (King and Rofe, 1976).

SEX CHROMOSOME EVOLUTION IN *Gehyra* AND *Heteronotia*

Detailed studies of intraspecific cytogenetic variation in *Gehyra* (King, 1979, 1983, 1984; Moritz, 1984a, 1986) and *Heteronotia* (Moritz, 1984b, Moritz et al., 1990) have revealed highly labile sex chromosome systems. Within each genus there is extensive variation among populations and there have been multiple origins of HSC's.

Three cytologically distinct HSC systems have been reported for *Gehyra*. A central Australian population of *G. nana* had a W chromosome marked by an additional site of ribosomal RNA genes, expressed as a novel heterozygous secondary constriction on pair 7 (Moritz, 1986). The NOR-bearing W was not fixed in the population; some females (9/49) lacked the additional NOR site while, curiously, a few (4/38) males were also heterozygous for the novel NOR. This variant was absent in populations 100 km to the south and 250 km to the north. Presumably, the addition of ribosomal genes does not represent a fundamental step in the differentiation of an HSC as this seems unlikely to suppress recombination or to effect other molecular changes. Instead, the novel NOR can be viewed as a cytological marker linked to a major sex-determining region. The few anomalies in its distribution among females and males may be due to limited recombination between the NOR and the putative sex-determining region. Alternatively, they may reflect environmental effects superimposed on a system of genetic sex determination.

In *Gehyra australis* all females from two populations were heterozygous for the addition of a heterochromatic short arm to the largest pair of chromosomes (King, 1977). This variant was not found in adjacent populations of the same species (King, 1983).

In a third species, *Gehyra purpurascens*, Moritz (1984a) reported extensive sex chromosome variation involving the largest pair of chromosomes. Here, all females had HSC's, but there were six different types of W chromosome. A detailed analysis employing N-, C- and G-banding indicated at least two independent derivations of W chromosomes from homomorphic elements, each involving multiple pericentric and paracentric inversions and, in one case, a centric transposition.

The situation in *Gehyra* is of particular interest because of the diverse mechanisms of differentiation involved. It is also remarkable that the HSC's are restricted to specific populations or species. Finally, the chromosomal location of the major sex determining region appears to be conserved. The chromosomes of pair 7 are the sex chromosomes in *G. nana* and the ZW pair of G. purpurascens was formed by tandem fusion of pairs 7 and 8 from the ancestral 2n = 44 karyotype (Moritz, 1984a, 1986). The derivation of the ZW pair in G. australis is not as clear, but the presence of an interstitial C-band and the probable origin of this pair by tandem fusion (King, 1983) indicate that the ancestral pair 7 may also be involved here.

The genus Heteronotia currently includes three species (Kluge, 1963; Storr, 1989) one of which, H. binoei, is clearly composite, including several diploid bisexual chromosome races and triploid parthenogenetic lineages derived by hybridization between them (Moritz, 1983; Moritz et al., 1990). Among diploids, the fourth pair of chromosomes is highly polymorphic for morphology and C-banding pattern relative to the other pairs. Most of these variants are characterized by shifts of terminal heterochromatin towards the centromere (Figure 1) and these morphs only occur in

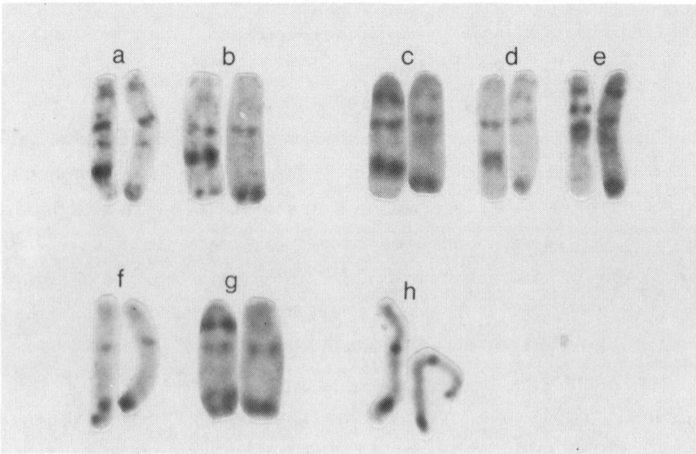

Figure 1. Comparison of W (left) and Z (right) chromosomes of Heteronotia binoei after C-banding. The letters above each pair identify the variants referred to in Table II.

heterozygous females (Table II). In a large sample of the CA6 race from the MacDonnell Ranges, all 18 females were heterozygous for one such variant whereas all 20 males were homozygous for the alternative morph (Table II).

This constitutes strong evidence for a ZW system involving chromosome pair 4. The sample sizes from other localities with variants of this type (Wb - Wh, Figure 1) are much smaller, but their distribution among sexes (Table II) suggests that these too represent ZW heteromorphisms. The HSC's only occur in three of the six chromosome forms (karyomorphs) of _H. binoei_, and then only in some populations (Table II).

TABLE II

Distribution of W chromosome variants among populations of _Heteronotia binoei_. The variants are those shown in Figure 1.

Locality	Chromosome Race	Number C-banded Male	Number C-banded Female	W chromosome
MacDonnell Ranges N.T.	CA6	20	18	Wa
Kulgera N.T.	CA6	0	2	none
100km N. Alice Springs N.T.	CA6	1	1	none
Cordillo Downs Stn S.A.	CA6	1	1	Wb
Wooramel Stn W.A.	CA6	1	4	none
Carbla Stn W.A.	CA6	0	3	none
Baroalba N.T.	NA6	1	2	Wc
Bing Bong Stn N.T.	NA6	1	2	Wd
Hayes Creek N.T.	NA6	0	1	Wc
Mataranka N.T.	NA6	2	2	Wc,We
Darwin N.T.	NA6	1	1	none
Wauchope N.T.	SM6	2	1	Wf
Turkey Creek W.A.	SM6	1	2	Wg
Tanumbirini Stn N.T.	SM6	0	1	Wh
Daly Waters Stn N.T.	SM6	1	1	Wh
Willeroo Stn N.T.	SM6	2	1	Wh
Halls Creek W.A.	SM6	0	1	none
Renner Springs N.T.	SM6	0	1	none
Enderby Island W.A.	SM6	0	1	none
Ayers Rock N.T.	SM6	1	1	none
Camooweal QLD	SM6	1	1	none
Karratha W.A.	SM6	1	1	none
Nanutarra Stn W.A.	SM6	1	1	none

A probable ZW heteromorphism has also been observed in some populations of an undescribed species (H. sp.) from central Australia (Moritz et al., 1990). Here, five females from two populations were heterozygous for a metacentric chromosome (possibly pair 4 or 5) which was absent from seven males. HSC's are absent from H. spelea and for the third species, H. planiceps, there are insufficient data.

A phylogenetic interpretation of these data (Figure 2) confirms that the ancestral condition is to have homomorphic sex chromosomes and that cytologically distinguishable sex chromosomes have arisen on several occasions. Even within a race, W chromosomes may have become differentiated more than once. For example, the interstitial C-bands of the Wa and Wb variants are more similar to those of the local Z chromosome variant they are to each other (Figure 1). An exception may be the Wf and Wg variants

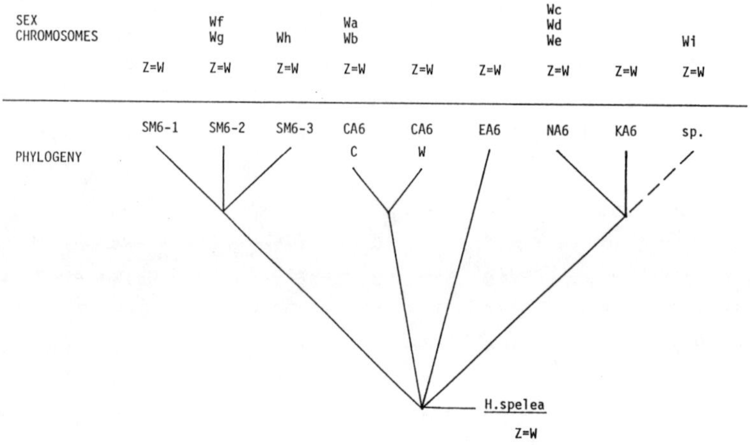

Figure 2. Distribution of the W chromosome variants in relation to the phylogeny of the various chromosome forms of Heteronotia binoei and H. sp. H. spelea was used as an outgroup. The placement of H. sp. as the sister group to the NA6-KA6 clade is tentative.

found in populations of the SM6 race from mid-northern Australia. These differ in the morphology and C-banding of the half of the chromosome proximal to the centromere, differences which also occur among local Z chromosomes. Thus, it is possible that the Wf and Wg variants are interconverted by recombination between the W and variants of the Z chromosomes (Figure 3).

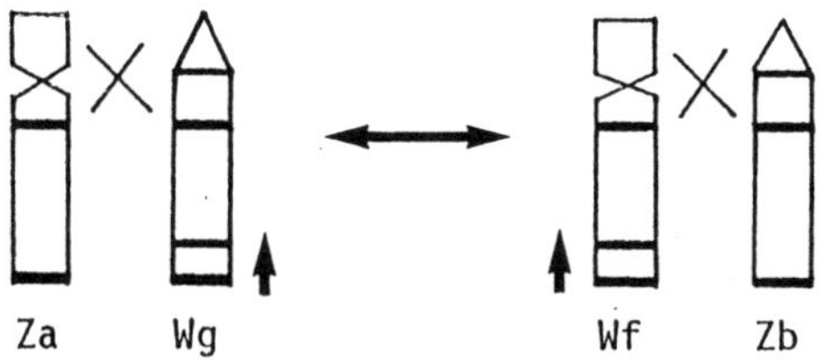

Figure 3. Diagram of possible interconversion between the Wg and Wf morphs by recombination between the W charomosomes and polymorphic variants of the Z chromosome.

The precise mechanism responsible for the differentiation of the W chromosomes of H. binoei is not clear. The observed shifts in C-band location could be due either to deletion and amplification of repeated sequences or to paracentric inversion. Given that no variants with both interstitial and telomeric blocks of heterochromatin have been identified, the latter mechanism seems more likely. G-banding has failed to provide any clearer resolution of the differences between Z and W chromosomes.

Comparison of the distribution of chromosome 4 variants among sexes suggests that the major sex (female) determining region of H. binoei is located on the half of chromosome 4 distal from the centromere, probably close to the telomere. This is the region encompassed by all of the sex-correlated shifts in heterochromatin. In contrast, polymorphism in the proximal half of chromosome 4, e.g., pericentric inversions ($p = 0.62$, $N = 46$ in SM6 populations) and pericentromeric C-bands ($p = 0.58$, $N = 30$ in western CA6 populations), show autosomal inheritance (Moritz, 1984b; Moritz et al., 1990).

The existance of polyploid H. binoei provides evidence that the W chromosome is dominant for sex determination in this species (Table III). All triploids are female and are thought to have originated by fertilization of unreduced ova produced by allodiploid females (Moritz, 1983). Their sex chromosome constitution would therefore be ZZW. The triploids examined by C-banding lacked the cytologically distinguishable W chromosome variants.

TABLE III

Predicted sex of different ploidy Heteronotia binoei for dominant vs. recessive W mmechanism of sex determination. The observations (in bold) fit the dominant W model.

Ploidy	Sex Chromosomes	Model	
		Dominant W	Recessive W
DIPLOID	ZW	**female**	female
TRIPLOID	ZZW	**female**	intersex?
TETRAPLOID	ZZZW	**female**	male

This is consistent with other evidence (Moritz et al., 1989; unpublished data) that they arose by hybridization between the western populations of the CA6 and SM6 races which lack HSCs (Table II). Tetraploids are occasionally formed by mating between the triploid parthenogenetic females and males; their sex chromosome constitution would be ZZZW. All four specimens of this type were female. Two had reduced gonads, but a third had apparently normal ovaries and appeared to have recently ovulated. These observations meet the expectations for a system of sex determination directed by a dominant, rather than a recessive, W chromosome (Table III). This contrasts with the situation in Lacerta, another lizard genus (family Lacertidae) with ZW sex determination and parthenogenesis. Here some hybrids between the diploid parthenogens and bisexual males are female, whereas others are males or intersexes, perhaps reflecting variation among species in the dominance relationships between Z and W chromosomes (Kupriyanova, 1989; Darevsky et al., 1989).

THE EVOLUTION OF SEX DETERMINATION

At least in theory, there are no profound barriers to the transitions between male and female heterogamety and between genetic and environmental sex determination (reviewed in Bull, 1983). For example, male heterogamety can evolve from female heterogamety via a multigenic system if there is appropriate genetic variation and some selective advantage to the new system. ESD can evolve from GSD if there is some environmental influence on sex and if there is variation among local environments for the ratio of male to female fitness. Alternatively, Deering and Ferguson (1988) suggest that GSD may evolve from ESD in response to increased uniformity of incubation conditions (e.g., with viviparity). The presence of strongly differentiated sex chromosomes is negatively correlated with diversity of sex determining mechansisms, either because a degenerate Y (or W) blocks some modes of transition between sex determining modes, or because changes in the sex determining mechanism may eliminate pre-existing HSCs (Bull, 1983; Hillis and Green, 1989).

Geckos illustrate all of the diversity needed to test the above models. Although the data are scant, effects of temperature on sex determination (i.e., ESD) and HSCs have been reported for the same species (*Gekko japonicus*), and both male and female heterogamety are observed in the family. The variability of the sex determining system and the concomitant rarity of highly degenerate sex chromosomes is consistent with the pattern in some other groups of reptiles (e.g., turtles, crocodiles, lizards; Bull, 1983; Olmo, 1986). The challenge now is to exploit this variation among geckos using ecological, genetic, and molecular approaches. For example, Bull et al. (1988) found that DNA from male and female *Eublepharis macularius* had similar patterns of hybridization to a probe for the putative testis determining gene of mammals (Page et al., 1987), suggesting that in this species the sequence is autosomal.

MECHANISMS OF SEX CHROMOSOME DIFFERENTIATION.

The early stages in the differentiation of sex chromosomes can be evaluated by comparing the properties of the relevant chromosome pair in closely related taxa with and without HSCs. At least for geckos, it appears that structural rearrangements typically preceed the invasion of euchromatin by heterochromatin and also precede allocyclic condensation, at least for mitotic chromosomes. In many cases, the changes are of the type expected to suppress local recombination (e.g., inversions and heterochromatin amplification) in accordance with the view that these function to alter linkage relationships (Ohno, 1967; Bull, 1983). However, for the geckos there are no direct comparisons of the distribution of chiasmata in the sex chromosomes with and without structural differentiation. Thus, it is not known if the rearrangements are a cause or a consequence of the suppression of recombination.

The observation that the primary step in the differentiation of gecko sex chromosomes is usually structural changes does not preclude alternative pathways in other groups. Indeed, there is no theoretical reason why there should one cytological mechanism responsible for the initial suppression of recombination. In some amphibian (Schmid, 1983), snake (Singh et al., 1976), and lizard (Olmo et al., 1987) species, recently derived (i.e., phylogenetically isolated) sex chromosomes differ in the proportion of heterochromatin, but not in size or shape. Although the banding data are seldom adequate to eliminate the possibility of modifications such as paracentric inversions, such cases suggest that invasion/amplification of heterochromatin may be the first step in differentiation. A particularly interesting situation was reported for the skink *Carlia longipes* by Donnellan (1985). Populations in the northern part of the range (Cape York, Australia) have an XY heteromorphism where the Y is much larger than the X and is almost entirely heterochromatic. In contrast, in a southern population had similar the X and Y were similar in size and shape and differed in that the Y had less centromeric heterochromatin than the X. The key observation is that the sex bivalent formed a ring in males with both types of HSC; apparently the suppression of recombination did not depend on an increase in Y chromosome heterochromatin (S. Donnellan, pers. commun.).

Whether or not the structural rearrangements or changes in amount of heterochromatin cause changes in recombination, the data are consistent with the view that the modification of recombination between the sex chromosomes is paramount. There are no indications of changes in the pattern of chromosome condensation (at least for mitosis; futher evidence is needed on allocycly in meiosis in systems with female heterogamety, e.g., Becak and Becak, 1981), which the "hijack" model (Jones, 1983, 1984) predicts should precede the suppression of recombination. The one possible example is the variation in mitotic replication banding reported in *L. lugubris* (Voloboeuv and Pasteur, 1988), but it remains to be demonstrated that this is a sex chromosome system.

A related problem highlighted by the analysis of gecko sex chromosomes relates to the extraordinary variability of the heteromorphic element. Why in *Gehyra* and *Heteronotia* does the W chromosome show so much variation compared to the Z or the autosomes? Possible explanations include (i) a higher rate of evolution because of the smaller effective population size and increased exposure to some forms of selection (Charlesworth et al., 1987), (ii) a higher probability of survival of rearrangements because of the lack of recombination between the HSCs (Voronstov et al., 1980), (iii) enhanced fixation because of the creation of novel adaptive linkage groups (Ohno, 1967; Bull, 1983), or (iv) an increased mutation rate, perhaps associated with the primary sex determining region (Moritz, 1984a). None of these were particularly compelling as explanations of the variation in *Gehyra purpurascens* (Moritz, 1984a).

PROSPECTS

The analysis of sex determination and of sex chromosome systems in geckos is in its infancy, but holds great promise. Many more analyses of intra- and interspecific chromosome variation using an array of banding methods are needed before the distribution and diversity of HSCs can be properly evaluated. Ultimately, it may be possible to place this information onto a phylogeny for the group to estimate ancestral states and to identify any phylogenetic constraints on shifts in sex determining mechanisms (e.g., Hillis and Green, 1989). Tests for ESD and comparisons of the chromosomal location and expression of candidate sex determining genes among closely related species with and without HSCs should provide information on the mechanism and evolution of sex determination.

ACKNOWLEDGEMENTS.

Thanks to Steve Donnellan and Max King for access to unpublished results, and to Jim Bull, Steve Donnellan, Dave Hillis, Max King and Greg Mengden for discussion or comments on the manuscript. This research was supported by grants from the Australian Research Council and the National Geographic Society.

REFERENCES

Becak, M.L., and Becak, W. (1981) Chromosoma 83, 289-293.
Becak, W. (1983) Differentiation 23, S3-S12.
Becak, W., and Becak, M.L. (1969) Cytogenetics 8, 247-262
Bull, J.J. (1983) Evolution of Sex Determining Mechanisms. Benjamin Cummings, London, Amsterdam.
Bull, J.J. (1987) J. Exp. Zool. 241, 143-148.

Bull, J.J., and Charnov, E.L. (1985) Evolution 39, 1149-1155.
Bull, J.J., Hillis, D.M., and O'Steen, S. (1988) Science 242, 567-569.
Charlesworth, B. (1978) Proc. Natl. Acad. Sci. (USA) 75, 5618-5622.
Charlesworth, B., Coyne, J.A., and Barton, N.H. (1987) Amer. Nat. 130, 113-146.
Darlington, C.D. (1958). The Evolution of Genetic Systems (2nd Ed.) Basic Books, New York.
Darevsky, I.S., Danielyan, F.D., Sokolova, T.M., and Rozonov, Y.M. (1989) In: Evolution and Ecology of Unisexual Vertebrates (R.M. Dawley and J.P. Bogart, Eds), New York State Museum Press, Albany, pp. 228-235.
Deeming, D.C., and M.W.J. Ferguson. (1988) Phil. Trans. Roy. Soc. Lond. B 322, 19-39.
Donnellan, S.D. 1985. Evolution of Sex Chromosomes in Scincid Lizards. PhD Thesis, Macquarie Univ., Sydney.
Hillis, D.M., and Green, D.M. (1989) J. Evol. Biol., in press.
Jones, K.W. (1983) Differentiation 23, S56-S59.
Jones, K.W. (1984) Chromosomes Today 8,242-255.
Jones, K.W., and Singh, L. (1985) Trends Genetics 1, 55-60.
King, M. (1977) In: Evolution and Reproduction (J. Calaby and H. Tyndale Biscoe, Eds), Aust. Acad. Sci, Canberra, pp. 55-60.
King, M. (1979) Aust. J. Zool. 27, 373-393.
King. M. (1981) In: Evolution and Speciation (W.R. Atchley and D. Woodruff, Eds), Cambridge Univ. Press, Cambridge, pp. 262-285.
King, M. (1983) Aust. J. Zool. 31, 723-741.
King, M. (1987) Aust. J. Zool. 35, 507-531.
King, M., and Rofe, R. (1976). Chromosoma 54, 75-87.
Kluge, A.G. (1963). J. Roy. Soc. West. Australia 46, 63-67.
Kluge, A.G. (1987). Misc. Pub. Mus. Zool., Univ. Mich.. (173), pp 1-54.
Kupriyanova, L.A. (1989). In: Evolution and Ecology of Unisexual Vertebrates (R.M. Dawley and J.P. Bogart, Eds), New York State Museum, Albany. pp 236-240.
MacGregor, H.C., and H.A. Horner (1980) Chromosoma 76, 111-122.
McBee, K., Bickham, J.W., and Dixon, J.R. (1987) J. Herpetol. 21, 68-71.
Mengden, G.A. (1981) Chromosoma 83, 275-287.
Moritz, C. (1983) Science 220, 735-737.
Moritz, C. (1984a) Chromosoma 90, 111-119.
Moritz, C. (1984b) Chromosoma 89, 151-162.
Moritz, C. (1986) Syst. Zool. 35, 46-67.
Moritz, C., Brown W.M., Densmore, L.D., Wright, J.W., Vyas, D., Donnellan, S., Adams, M., and Baverstock, P. (1989) In: Evolution and Ecology of Unisexual Vertebrates (R.M. Dawley and J.P, Bogart, Eds), New York State Museum press, Albany, pp 87-112.
Moritz, C., Adams, M, Donnellan, S., and Baverstock, P. (1990) Copeia, in press.
Ohno, S. (1967) Sex Chromosomes and Sex Linked Genes, Springer, Berlin, Heidelberg, New York.
Olmo, E. (1986) In: Animal Cytogenetics 4 Chordata 3 (B. John, Ed), Gebruder Borntraeger, Berlin-Stuttgart.
Olmo, E., Odierna, G., and Capriglione, T. (1987). Chromosoma 96, 33038.
Page, D.C., Mosher, R., Simpson, E.M., Fisher, C., Mardon, G., Pollack, J., McGillivray, B., de la Chappelle, A., and Brown, L.G. (1987) Cell 51, 1091-1104.
Ray-Chaudhuri, S.P., Singh, L., and Sharma, T. (1971) Chromosoma 33. 239-251.
Schmid, M. (1983) Differentiation 23, S13-S22.
Singh, L., Purdom, I.F., and Jones, K.W. (1976) Chromosoma 59, 43-62.
Singh, L., Purdom, I.F., and Jones, K.W. (1980) Chromosoma 79, 137-157.

Solleder, E., and Schmid, M. (1984) Amphibia-Reptilia 5, 339-345.
Storr, G.M. (1989) Rec. West. Australian Mus. 14, 269-274.
Tokunaga, S. (1985) Develop. Growth Diff. 27, 117-120.
Volobouev, V., and Pasteur, G. (1988) Heredity 60, 463-467.
Voronstov, N.N., Lyapunova, E.A., Borissov, Yu. M., and Dougal, V.E. (1980) Genetica 52/35, 361-372.
Wagner, E. (1980) Quart. Rev. Biol. 55, 21.
White, M.J.D. (1973) Animal Cytogenetics and Evolution (3rd Ed). Cambridge Univ. Press, Cambridge.
Yoshida, M.C., and Msahiro, I. (1974). Chrom. Inf. Serv. 17, 29-31.

Cytogenetics of Amphibians and Reptiles
Advances in Life Sciences

THE W CHROMOSOME DURING THE EVOLUTION AND IN SEX ABNORMALITIES OS SNAKES. DNA CONTENT, C-BANDING.

M. Luiza Beçak, M. Nazareth Rabello-Gay, W. Beçak, M. Soma, Radenka F. Batistic and Isa Trajtengertz.

Serviço de Genética do Instituto Butantan, Av.Vital Brazil,1500, São Paulo, SP, Brasil.

SUMMARY: We report that: a. evaluation of DNA content in cells of snakes from three families, Boidae, Viperidae and Colubridae showed variation in the last; b. the W-sex chromosome is polymorphic in size and C banding in *B. insularis*, a viperid consisting of males, unfertile females and fertile females but with hemipenis; c. polymorphism of the W also occurs in *B. jararacussu*, a normal related species from the continent. We propose that: 1. the variation of DNA content points to polyphyletic diversification in colubrids; 2. the abnormal sex phenotypes and the altered fertility in *B. insularis* females are related to chromosomal rearrangements probably by insertion sequences in the W. The levels of "intersexuality" in *B. insularis* females would be related to the degree of W alteration. Unfertile females without hemipenis (10%) would result from virilized females through W reversion.

Most snakes have ZZ/ZW chromosomal sex-determining mechanism in which females are the heterogametic sex. ZW heteromorphic sex chromosomes (s.c.) were first reported in two poisonous snakes: *Bothrops jararaca* (Beçak, W. et al, 1962) and *Vipera berus* (Kobel, 1962). Comparative cytogenetical studies showed that the s. c. in Brazilian snakes present different levels of morphological differentiation apparently coincident with the degree of evolution of the families. The primitive family Boidae has homomorphic s. c. and the highly evolved Viperidae has well differentiated s. c. regarding length and centromere position.

In Colubridae s. c. are homomorphic or heteromorphic, differing either by centromere position or by centromere position and size. In most of these species the W is smaller than the Z which corresponds to the 4th pair in the genome. In some Colubridae species the W is exceptionally the largest in the genome, probably due to eventual duplications as in Clelia occipitolutea, Clelia c. plumbea, Rachidelus brazili and Pseudoboa nigra (Beçak, W., 1964, 1966, 1983; Beçak, W. et al, 1966; Beçak, W.and Beçak, M. L., 1969; Beçak, M. L. et al, 1971-1973). The occurrence of heteromorphic ZW chromosomes of equal size in Colubridae led Beçak, W. (1964) and Ohno et al (1967) to suggest that the first step for differentiation of the W was a pericentric inversion. This suggestion implies that the inhibition of crossing-over due to the inversion resulted in accumulation of sex determing genes in the W. According to Beçak, W. (1983) sex is determined by a major dominant gene or group of genes active in males or females. An eventual inversion inhibiting crossing-over may lead to differentiation (heterochromatinization) of the incipient W and further changes in size by deletion or duplication.
Autoradiographic studies of snake s. c. did not show in the homogametic sex inactivation of the Z, analogous to the female X-sex chromatin of mammals (Bianchi el al, 1969).These data also evidenced that the ZW pair finishes its replication earlier than the macroautosomes. A W with an early replicating pattern was especially observed in Philodryas patagoniensis (Colubridae).

Similar information regarding the evolution of s. c. was reported for Indian snakes of the families Boidae, Colubridae and Elapidae (Singh et al, 1968a, b, 1970). Several reports showed: 1. the presence of multiple s. c. in Elapidae and Hydrophiidae; 2. the occurrence of species with the W larger than the Z; 3. that, except for the Boidae, W condenses in a chromocentre (W-chromatin) in interphase nuclei; 4. that W is C band positive and allocyclic with early or late replicating patterns (Singh et al, 1970, 1972a, b, Ray-Chaudhuri et al,1971, Ray-Chaudhuri and Singh, 1972, Singh, 1974). The presence of heterochromatin in the W of colubrid snakes with different levels of s.c. differentiation led the authors (Ray-Chaudhury et al, 1971) to suggest that heterochromatinization of the W rather than pericentric inversion may be the first stage in the differentiation of s. c. in snakes. The use of fluorescent staining techniques showed enhancement of the W chromatin in colubrid and viperid species. Yet, the W-sex chromatin body was not observed in the boid Boa c. amarali in which s. c. are undifferentiated. The presence of the bright fluorescent material in interphase nuclei of snakes with heteromorphic s. c. and its absence in somatic metaphases of the same species seem to reflect the functional stage of activity of the W DNA (Beçak, W. and Beçak, M. L., 1972).

Further studies evidenced that C-bands are not homogeneous in the W of snakes and that heteromorphic s. c. also occur in the primitive snake Acrantophis dumereli (Boidae) (Mengden and Stock, 1980, Mengden, 1981). These facts argue in favour of the hypothesis that a pericentric inversion may be the first step in the differentiation of the s. c.. Nevertheless, the findings in Philodryas chamissonis (Moreno et al, 1987) support the hypothesis that heterochromatinization is the first step in the W isolation. In this species the ZW pair was found to be isomorphic. The W is totally C-band positive appearing as W-chromatin in interphase nuclei. The genus Philodryas includes several species with well differenciated s. c. (Beçak, W., 1964).

Molecular studies have indicated that the different patterns of DNA replication exhibited by the W result from satellite DNA sequences present in this chromosome. These satellite DNAs were isolated by analytical ultracentrifugation from Elaphe radiata (Sat III and IV) and the banded krait Bungarus fasciatus (minor satellite or Bkm). The Bkm sequence is similar to sat IV. In situ hybridization of Bkm demonstrated that this DNA sequence maps to the W of most ophidians with different levels of s. c. differentiation. Molecular hybridization of Bkm to male and female snake DNA on nitrocellulose filters demonstrated that this satellite DNA is universally present in snakes. In the primitive Boidae both sexes are positive although showing fewer copies. The presence of Bkm in the W was considered to be the first step in the isolation process of s. c. (Singh et al, 1976, 1979a, b, 1980). The Bkm sequence, which consists of highly repetitive GATA-GACA tetranucleotides, was identified in the male determining region of the Y chromosome (TdY) of the mouse and in the proximal region of the Drosophila X (Singh et al, 1981, 1984, Singh and Jones, 1982). Besides the presence of Bkm in the human X this DNA sequence was shown in chromosomes 6 and 11 but not in the Y (Kiel-Metzger et al, 1985). A Bkm sequence was also detected in the horse Y in both normal XY males and XY sex-reversed females (Kent et al, 1988). The GATA repeats seem to encode mainly the amino acids leucine, serine, isoleucine and tyrosine (Singh et al, 1984, Jones and Singh, 1985). The real function of Bkm, i.e. whether it corresponds to the sex determination sequence, is still discussed. It is known that the Y of mammals is a male determining chromosome directing gonadal differentiation to the testis, the development of secondary sex traits being mediated by gonadal hormones. It has been proposed that the male-specific histocompatibility antigen (H-Y) which is controlled by a gene(s) on the Y is a testis inducer factor. H-Y antigen was found in all species having male heterogamety (XX, XY) and in animal groups with female heterogamety (ZZ, ZW). In the latter it was suggested that the H-Y (H-W) antigen induces the development of the ovary, the gene being located in the W (Wachtel et al, 1975). In snakes, H-W antigen was found in B. jararaca (Viperidae), Mastigodryas b. bifossatus and Waglerophis merremii (Colubridae) (Nakamura et al, 1987). The "H-Y hypothesis" was not confirmed by the recent demonstration that H-Y expression is not necessary for testis determination in mice (McLaren et al, 1984). The testis-determining gene in man (TDF) maps to the short arm of the Y while the gene controlling the expression of H-Y antigen (Hya) maps to the proximal part of the long arm; most of the rare XX men are negative for the H-Y antigen (Sympson et al, 1987).

In this paper we analysed the chromosomal rearrangements

related to the variaton of the W in the evolution and in abnormal sex phenotypes of Bothrops insularis. The data were obtained by cytophotometry, chromosome morphology and C and NOR banding techniques. B. insularis is a poisonous snake restricted to Queimada Grande Island (São Paulo, Brazil) , whose population consists of males, unfertile females and fertile females but with a bilateral or unilateral hemipenis. The "intersexes" were classified as genetical females based on the analysis of somatosexual characters and the anatomy of the gonads. Sex-ratio was determined with two samples of specimens captured between 1914 and 1921, and from 1946 to 1953 with a gap of nearly 25 years. The analysis of 367 specimens (154 males, 176 "intersexes" and 37 females) showed a reduction of the number of males, an increase of the "intersexes"and a stationary frequency of females (10%). One of the hypotheses raised to explain the sexual abnormalities was the occurrence of mutations that might produce "intersexuality" (Hoge et al, 1959). The karyotype of B. insularis was first described in an "intersex" specimen as normal regarding the presence of a W apparently comparable to those found in continental snakes (Beçak,W.,1964). This fact confirmed the suggestion by Hoge et al (1959) that "intersexes" are genetically females. Further cytogenetical studies revealed the occurrence of a polymorphic W chromosome in this species. In ZW intersexes, the W was found to be larger than that of unfertile females. The difference in length was explained to be caused by a translocation probably from the Z to the W. The translocated segment would contain male determining genes (Beçak, M. L. et al, 1970).These preliminary cytogenetical data on the polymorphism of the W are included here together with new findings obtained by C-banding using specimens captured during the period from 1984 to 1988. A similar polymorphism of the W was found in B. jararacussu, a related species from the continent.

MATERIAL AND METHODS

The total DNA amount was measured in 13 species of snakes: Epicrates cenchria poecilogyrus (Boidae, from Ipomeri, Goiás, female); Rachidelus brazili (from Guararapes, São Paulo, female); Pseudoboa nigra (from São José do Rio Preto, São Paulo, male); Clelia occipitolutea (from Porto Feliz, São Paulo, male); Sordellina punctata (from Miracatu, São Paulo, female); Helicops modesta (from Jarinu, São Paulo, female); Chironius foveatus (from Jarinu, São Paulo, female); Leimadophis p. schottii (from Guaraçai, Minas Gerais, male); Elaphe quadrivirgata (from Japan, male) (Colubridae) and Bothrops moojeni (from Uberaba, Minas Gerais, male); B. neuwiedi (from Itaipu, Paraná, male); B. insularis (from Ilha da Queimada Grande, São Paulo, "intersex"); Agkistrodon h. blomhofii (from Japan, male) (Viperidae). DNA values were estimated by using the Feulgen reaction with scanning cytophometry. Blood was withdrawn by heart or vein puncture in EDTA (1%) at a ratio of 1mg EDTA/ml of blood. The preparations were fixed with methanol and stored at room temperature until staining. For each animal we plotted the hydrolysis curve. DNA was determined on erythrocyte nuclei, using λ =540 nm in a Zeiss cytophotometer. In order to reduce the effect of any possible technical and instrumental error, preparations of Bufo ictericus were always stained and measured together as controls. The absolute DNA content of B. ictericus has been previously determined using the known values for human lymphocytes and kidney cells of Rattus norvegicus.

The cytogenetical study of B. insularis was carried out on 21 specimens collected during two different periods (from 1968 to 1970 and from 1984 to 1988). The morphological analysis was done on 6 ♂, 3 ♀ and 8 ⚥, one female with a dubious sex diagnosis had only one ovary. C-banding was performed on 1 male and 3 "intersexes" from the second sample. NOR banding was analysed in one of those "intersexes". The chromosomes were characterized and paired according to relative length, arm ratio and centromere index, as previously described (Beçak, W., 1964).

Chromosomes were obtained by the standard squash technique from different tissues (intestine, testicle, ovary and spleen) as well as by short term lymphocyte culture. C banding was obtained by the method of Sumner (1972). NOR banding was performed according to the technique described by Howell and Black (1980).

Chromosomes of B. jararacussu were studied with the squash technique in one female collected at Itanhaem (year 1985) a continental locality straight in front of Queimada Grande Island. Phenotype abnormalities were not reported for this species. The specimen analysed seems to be normal and had an offspring of 54 animals.

RESULTS

Cytophotometric data show that DNA content is different among Boidae, Colubridae and Viperidae.In the colubrids, the amount of total DNA falls into two groups. The highest values comprise six species (Rachidellus brazili, 2n=50, 5.39pg, Pseudoboa nigra, 2n=50, 4.95pg, Clelia occipitolutea, 2n=50, 4.90pg, Sordellina punctata, 2n=52, 4.83pg, Helicops modesta, 2n=34, 5.03pg and Chironius foveatus, 2n=34, 4.54pg). The lowest values include Leimadophis p. schottii (2n=32, 3.84pg) and Elaphe quadrivirgata (2n=36, 2.63pg). In the viperid species (2n=36), DNA amount was more uniform: Bothrops moojeni, B. neuwiedi, B. insularis and Agkistrodon h. blomhoffii (2.17, 2.45, 3.82 and 2.55 pg , respectively). In Epicrates c. poecilogyrus,the only boid(2n=36) analysed DNA content was found to be 3.06pg.

The male and female karyotypes of the viperids reported here were previously described by us with the exception of A. h. blomhoffii. The female karyotype we analysed is comparable to those of other viperids with 2n=36, the W being the fourth in the genome. The karyotypes of the colubrids with a high 2n number (2n=50) were also reported with the exception of S. punctata. This species has 2n=52, an uncommon number in colubrids. The karyotype consists of acrocentric macrochromosomes, the W being the largest in the complement. As in other colubrid species (2n=50), this is a primitive karyotype. The Z and W chromosomes are submetacentrics, except in C. occipitolutea where the Z is acrocentric. The colubrids with a lower 2n number have already been studied with the exception of H. modesta and E. quadrivirgata females.

Table I: Macrochromosome measurements of B.insularis (R.l.: relative length; A.r.: arm ratio; C.i.: centrome index)

chromosome pairs	1	2	3	4 Z	4 W	5	6	7	8
R. l.	27.37	21.77	14.66	9.21	6.66	7.74	7.39	6.24	5.39
A. r.	1.12	1.75	1.15	1.11	1.80	2.44	2.76	1.97	2.45
C. i. ♀	0.47	0.36	0.47	0.47	0.37	0.31	0.28	0.35	0.30
R. l.	25.82	21.17	15.41	9.78	8.60	8.12	6.84	6.51	5.84
A. r.	1.28	1.74	1.24	1.31	1.87	2.44	3.36	1.96	2.83
C. i. ⚥	0.47	0.37	0.45	0.44	0.36	0.30	0.24	0.33	0.27

The cytogenetical data show that the specimens of B.insularis grouped as males and unfertile females without a hemipenis (♀,Hp-) have apparently normal karyotypes comparable to those of related continental snakes (2n=36, 14M+20m, ZZ/ZW). A remarkable fact observed in the fertile females with a hemipenis (⚥, Hp+, "intersexes") was the presence of a large W (Table I). Large and small W chromosomes were designated as W_2 and W_1, respectively. The study of about 30 metaphases of each "intersex" revealed that the W_2 may exhibit a peculiar region apparently located in the median part of the long arm. At this site both chromatids are closely arranged and appear to be joined (Figs.1a,1b and 2a). Yet, W_2 can appear smaller in some nuclei of the same "intersex". In this case it is apparently comparable to the W_1 of unfertile females and to the W previously described in one "intersex" by Beçak, W. (1964). The intriguing presence of both ZW_1 and ZW_2 nuclei was observed also in the female with dubious sex diagnosis (Figs. 1c and 1d). In this specimen and in the

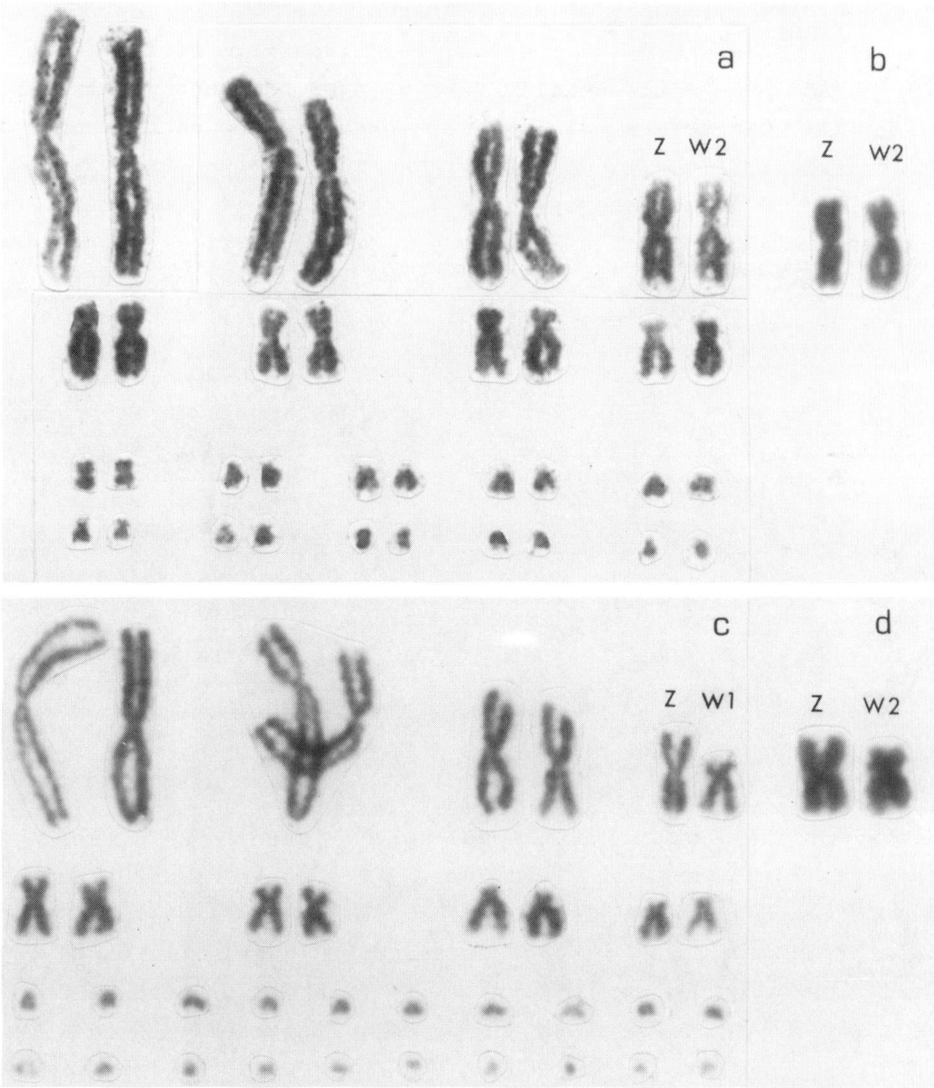

Figure 1. W_2-type chromosome found in lymphocytes from an "intersex" specimen (a, b). This chromosome may appear with closely arranged chromatids in the median region of the long arm (a). W_1 an W_2-types in spleen cells from a dubious female (c, d). Giemsa stain.

"intersexes" a fragment comparable in size to a microchromosome was also observed. The fragment is easily confused with a microchromosome that usually shows precocious centromere disjunction. In some nuclei we also observed the occurrence of an acentric W (Fig. 2). The frequency of W_2 and W_1 was determined in 10 metaphases from each specimen (4 ⚥ and one sex dubious female). The frequency of W_2 varied among the "intersexes" (from 70% to 90%). The dubious female exhibited W_1 in 60% of the karyotypes.

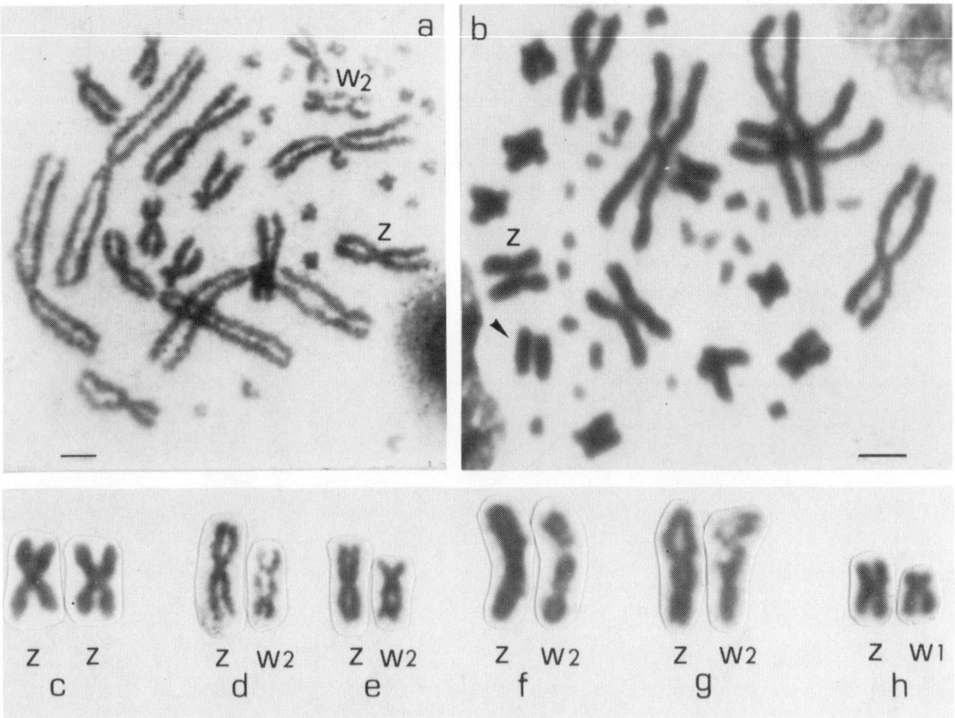

Figure 2. Abnormalities of the W_2 in two "intersex" specimens (a - d, e and f, g, respectively); in a dubious female (b) and in a normal female (h). Male ZZ pair of a spleen cell is included for comparison (c). Note: chromatids apparently joined at the long arm (a - d and e; lymphocytes and f, g, ovary); acentric aspect (arrow in b; spleen) and W_1-type reduced size(h, intestine). One micro is frequently lacking in ZW_2 cells (a). Giemsa stain, bars = 2 µm.

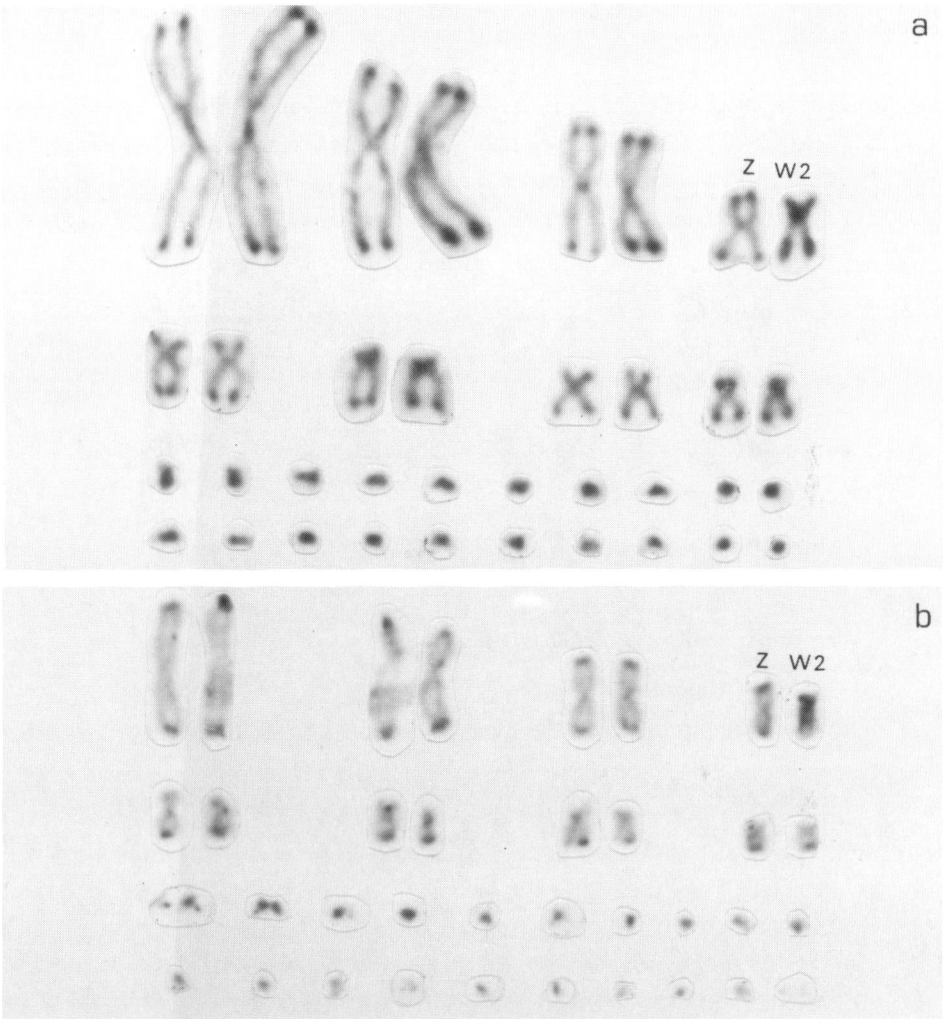

Figure 3. Karyotypes of two "intersexes" (a,b) showing different C-banding patterns on the long arm of the W_2 (intestine).

Chromosome C banding in one male and three "intersexes" revealed the presence of telomeric and centromeric heterochromatin in all macroautosomes and in the Z, with the microchromosomes being entirely heterochromatic. The W_2 appeared to be variegated with positive and negative C-bands. This is an intriguing fact in snakes with a differentiated W where this chromosome is entirely heterochromatic, as described by Singh et al (1976). In the "intersexes" examined the short arm of the W_2 is heterochromatic but the C banding pattern differs in the long arm of the W_2. In one specimen the positive band is telomeric and in another it is interstitially located (Figs. 3a and 3b). The latter specimen also presented the small W_1-type which is

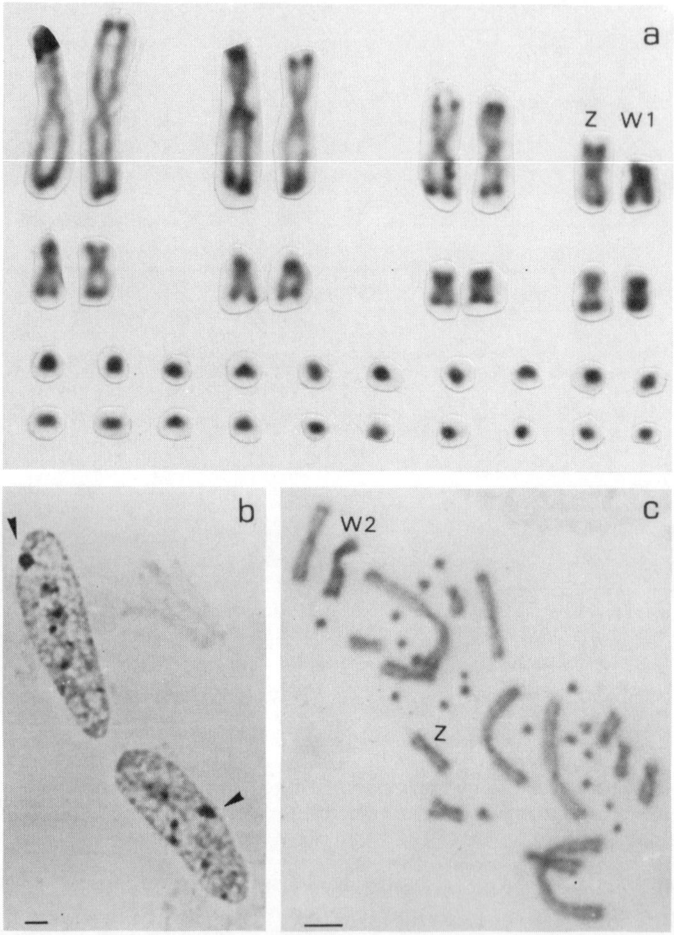

Figure 4. a. Heterochromatic W_1-type of the same "intersex" shown in Fig. 3b (intestine cell). b. W-sex chromatin in intestine interphase nuclei of B. jararacussu (arrowheads). c. polymorphic W_2 of B. jararacussu in an intestine cell metaphase. C banding; bars = 2 µm.

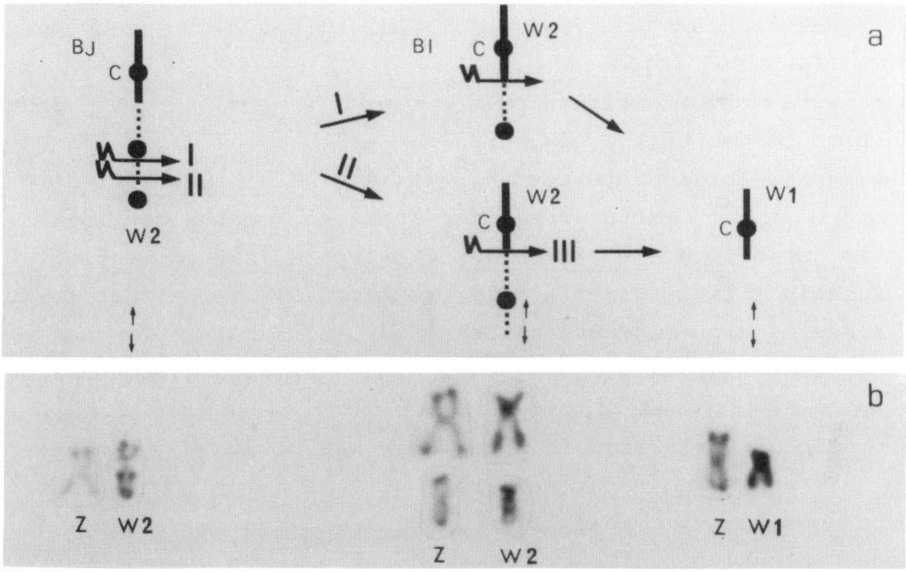

Figure 5. A model to explain diversification in Bothrops and the origin of the intersexuality in B.insularis (a); based on the evidence of polymorphic W_2/W_1 chromosomes by morphology and C banding studies (b). We propose that the polymorphic W_2 of both B.jararacussu (B.j.) and B.insularis (B.i.) were formed by translocations (probably of IS-type), followed in B.i. by other chromosome alterations (breaks or u.s.c.e.). In B.i. these changes may have involved the gene(s) that normally inhibits the development of the hemipenis in female embryos. Accordingly, a partial repression of the AHpD gene (antihemipenis development) that normally leads to the regression of the medullary cords in an imature ovary may cause the appearance of a more or less developed hemipenis in some females (ZW_2 , Hp+) . The activity of this gene would be restored in revertant females without hemipenis (ZW_1, Hp-) via W_2/W_1 reversion. This proposition agrees with previous suggestion that the masculinization of the cloacal territory in some females would be caused by the eventual persistence of the medullary cords for a longer time in an imature gonad (Hoge et al, 1959). Arrows (I, II and III) = presumed sites of chromosomes changes; dark lines and spheres = heterochromatin, C = centromere.

totally heterochromatic (Fig. 4a). This W_1 was also evidenced in the remaining "intersex".

NOR-bands were identified on the short arm of four microchromosomes (Fig. 5, see Discussion).

The particular aspects found in the W of B. insularis led us to compare it with those of related species from the continent. For this purpose, a B. jararacussu female was studied. Surprisingly this specimen also presented a large W_2 different from those described by Beçak, W. (1964). It has a heterochromatic short arm, with the long arm being euchromatic exception for the telomere and an interstitial region (Fig. 4c). W-chromatin was evidenced in interphase nuclei from intestine cells in C banding preparations (Fig. 4b) This sexual chromocentre was not conspicuously identified in B. insularis. NOR-bands were detected in one pair of microchromosomes.

DISCUSSION

The principal findings reported here are: the variation of total DNA content in snakes; b) the occurrence of a polymorphic W in abnormal sex phenotypes of B. insularis, an isolated Viperidae species from Queimada Grande Island; c) the presence of a polymorphic W in B. jararacussu, a normal species from the continent.

Previous findings have shown that snakes are a uniform group regarding karyotype morphology, total DNA content and Z size (Beçak, W., 1964). This suggestion was based on the following facts: 1. the majority of snakes have a basic karyotype of 2n=36 consisting of 8 pairs of macrochromosomes and 10 pairs of microchromosomes; 2. the amount of genomic DNA does not vary substantially; 3. the Z comprises 10% of genomic DNA. Deviations of karyotype number and morphology were evidenced in Colubridae which include several species with primitive karyotypes and different levels of heteromorphism. The existence of a heteromorphic ZW of equal size in Colubridae indicated that the first step in the isolation of the W may be a pericentric inversion (Beçak, W., 1964, Ohno et al, 1967, Beçak, W., 1983). Alternatively, it was postulated that one of the primary mechanisms of sex differentiation is the accumulation in the W of the highly repeated Bkm satellite DNA followed by heterochromatinization and further alteration in the morphology (Ray-Chaudhuri et al, 1971, Singh et al, 1976, 1980, 1981). In contrast to the conservation of DNA content of the Z, the relative length of the W varies between 4% and 27% in snakes (Singh et al, 1972a). The analysis of Bkm content in relation to the size of the W also indicated that other DNAs account for variation of the W. This seems to be the case for B. walliwal in which the W is entirely C-band positive but with a Bkm sequence only on the long arm (Singh et al, 1980). The description of polymorphic W chromosomes created by translocations and fissions in different populations of B. caeruleus (Elapidae) also suggests the presence of eventual insertion sequences (IS) altering this s. c. (Singh et al, 1979a). The findings that some Colubridae snakes with homomorphic s. c. have Bkm sequences suggested that crossing over was blocked even in homomorphic s. c. and that changes in the W DNA occurred rapidly and not over a long period of time as stated in Muller's classical model . According to this model, s. c. had diverged slowly via gradual accumulation of mutations. Instead, the "heterochromatinization hypothesis" for the W differentiation argues that accumulation of Bkm sequences is followed by rapid structural alterations . These would be caused by mutations involving the insertion of mobile genetic elements near the gene whose function is then modulated (Jones and Singh, 1985).

Our present data show that, in contrast to the uniformity of genome DNA content in viperids, the colubrids fall into two distinct groups: one group with a larger DNA amount (4.54 to 5.39pg) including species both with a high diploid number (2n=50, 52) and lower 2n number (2n=34); another group with a smaller DNA amount (2.63 to 3.84pg) has species with a lower 2n number (2n=36, 32). These findings seem to indicate the existence of polyphyletic lines of diversification in the colubrids. The differences in the karyotypes of the first group can be explained by centric fusions. Yet the differences in the W suggest that certain chromosome rearrangements (duplications, deficiencies) were favoured in this chromosome. This could mean the presence of DNA sequences with different susceptibility to certain rearrangements in the W. In primates too, some chromosomal rearrangements occurred more frequently in some families than in others (Dutrillaux et al, 1981). The uniformity of DNA content described here in Viperidae,a high evolved family, confirms previous data (Beçak, W., 1964) showing that this group of snakes has comparable DNA content and evolved karyotypes in which the W corresponds to the fourth pair.The highest value was found in B. insularis, a species with difficult phylogenetic position. In this species the cytogenetical data evidenced the presence of a polymorphic W differing in size and C banding patterns. The morphological findings obtained by conventional Giemsa staining revealed the presence of a large W_2 in 8 "intersexes" and a small W_1 in 2 females. Two crucial aspects were also observed among the "intersexes" and in one presumed female: a. the concomitant presence of both ZW_2 and ZW_1 nuclei in the same specimen, the frequency of ZW_2 being related to the level of masculinization; b. the presence of abnormal chromosome configurations (acentric fragments, acentric W, dicentric-like W). The large W_2 with variegated C banding patterns on the long arm was identified in two "intersexes". One of these specimens also presented ZW_1 nuclei in which W_1 was totally heterochromatic. This W_1 was also evidenced in a third "intersex".

The small W_1 observed in two ZW_1, Hp-, ♀ as well as in some cells of ZW_2, Hp+, ⚥, can be explained by normal condensation of the W_2. Nevertheless, we cannot eliminate the possibility that the smaller size of this chromosome is pointing to a deletion. This latter possibility is supported by the evidence of acentric fragments in some metaphases of the "intersexes".The peculiar aspect of some W_2 with apparently joined chromatids in the median part of the long arm seems to indicate a structural tendency of this chromosome. Nevertheless, we cannot eliminate the possibility that this is a dicentric chromosome. We should also consider the possibility that this uncommon W_2 may be an atelomeric W_2 frequently connected to the telomere of a microchromosome. This alternative interpretation is based on the evidence that the W_2 of one "intersex" does not show telomeric heterochromatin on the long arm. This explanation is also supported by the observation that one microchromosome is frequently missing in karyotypes having this particular dicentric-like W_2.

The presence of euchromatin in the large W_2 allow us to suggest that the increase in the size of this chromosome was probably caused by a translocation either from the Z or a macroautosome to the wild chromosome.This assumption is based on the suggestion that the W is heterochromatic in snakes with the exception of the boids (Ray-Chaudhuri et al, 1971).The different aspect of C banding found in the W_2 of two "intersexes" seems to indicate that this presumed translocated chromosome is of unstable morphology. The different location of the heterochromatin could be the result of a paracentric inversion. Accordingly, the abnormal chromosome aspects (acentric fragment, deficient W_1 and dicentric-like W) would be explained by gene recombination inside such paracentric inversion.Considering that chromosomal alterations also occur by gene recombination through unequal sister chromatid exchange (u.s.c.e.), which can be mediated by duplications/transposons (Tartof, 1988), we are also inclined to suggest this event to explain the alterations of the W. The interesting fact focalized by Hoge et al, 1959, that both "intersexes" and unfertile females are generated by the same mother ("intersex") led us to imagine that perhaps W_1 and W_2 could be produced by sister W_2 chromatid exchange. This event would occur in pre-meiotic germ cells or at meiosis.This implies that revertant ZW_1 females and ZW_2 "intersexes" are necessarily reciprocal products of the same genetic recombination in a same ZW_2 mother. In this case the frequency of unfertile females without a hemipenis might be a function of the frequency of W_2/W_1 reversion. Such sugestion is in accordance with data on sex-ratio showing that the frequency of unfertile females is being maintained (5 to 10%) in the island (Hoge et al, 1959, Federsoni, Buononato and Puorto, pers. comm.). The u.s.c.e. hypothesis would explain the differences in size of W_1 and W_2 and the different C-banding patterns exhibited by W_2 chromosomes of two "intersexes".Nevertheless, this genetic event does not explain the occurrence of fragments and presumed dicentric W_2 in the "intersexes". Thus, we are inclined to consider that the chromosome alterations may involve the insertion of eventual mobile genetic elements (I.S.).

Polymorphic W chromosomes created by translocations and fissions were described in different populations of _B.caeruleus_. The presence of an eventual insertion sequence in the W was also suggested to explain the occurrence of multiple s. c. in snakes (Singh et al, 1979a). I.S. occurs in some organisms such as _E. coli_, _Drosophila_ and maize. In _Zea mays_, breakage of chromosome 9 followed by fusion of broken ends of sister chromatids leads to the production of a dicentric, an acentric fragment and a deficient chromosome (McClintock, 1951b in Lima-de-Faria, 1983). The uncommon chromosomal abnormalities found in "intersexes" with polymorphism of C banding allow us to speculate, whether the W could have a dissociation site as described in _Zea mays_ in the dissociation-activation interation system investigated by McClintock.In this case we wonder whether the activity of a gene(s) that normally inhibits the hemipenis development (AHpD, anti-hemipenis development) in female embryos could be partially repressed in virilized females due to the translocation in the W_2 chromosome. Accordingly, gene activity would be restored in unfertile ZW_1 females with the loss of the transposed gene through W_2/W_1 chromosomal reversion.Both u.s.c.e. and I.S. hypotheses infer that the phenotypic variegation revealed by the alteration of the hemipenis development and fertility in the females of _B. insularis_ would be a position effect that can be visualized at the chromosomal level.Our data showed that a polymorphic W also occur in _B. jararacussu_, as evidenced in one female from Itanhaem, a continental site in front of Queimada Grande. Other specimens are needed to trace the origin of these polymorphisms and phylogenetic relationships (Figs. 5a and 5b).

The location of NOR bands in the microchromosomes of *B. insularis* and *B. jararacussu* confirms the description of ribosomal genes in these chromosomes of *B. jararaca* by Beçak, W. (1964). In embryos of *B. alternatus* secondary constrictions at macrochromosome (pairs 5 and 6) were reported (Beçak, M. L. and Beçak, W., 1973). Such secondary constrictions were not observed in adult specimens. We have proposed that at the early stage of development, where a higher synthesis rate is needed cytologically detectable secondary constrictions would be increased. In the Colubridae family the species *Philodryas chamissonis* has the NOR located in the long arm of pair 2. As secondary constrictions in other *Philodryas* species occur in pair 5 it was suggested that translocation of insertion type may have had a role creating intrageneric polymorphism (Moreno et al, 1987).

Acknowledgments: Supported by grants from CNPq and FAPESP.

REFERENCES

Beçak, M.L., Beçak, W., Roberts, F.L., Shoffner, R.N.,Volpe,E.P. (1971-1973) Chromosome atlas: Fish, amphibians, reptiles and birds, vols.1-2. Springer-Verlag, New York Heildelberg Berlin.

Beçak, M.L. and Rabello,M.N.(1969) Chromosoma (Berl.) 26,188-200.

Beçak, M.L., Rabello, M.N. and Beçak, W. (1970) Ciencia e Cultura E-76, 119 (abstract).

Beçak, W. (1964) In: Venomous animals and their venoms. Vol. 1 (Bucherl, W., Buckley, E., Deulofeu, W.,Eds.) Academic Press, New York, pp. 403-590.

Beçak, W. (1966) Mem. Inst. Butantan 33, 775-798.

Beçak, W. (1983) Differentiation 23(Supl.), 3-12.

Beçak, W., Beçak, M.L., and Nazareth, H.R.S.(1962) Cytogenetics I, 305-313.

Beçak, W., Beçak, M.L. and Nazareth, H.R.S. (1966) Mem. Inst. Butantan 33(1), 151-152.

Beçak, W. and Beçak,M.L.(1969) Cytogenet.Cell Genet.8,247-262.

Beçak, W. and Beçak, M.L. (1972) Experientia 28, 228-229.

Bianchi, N.O., Beçak, W., de Bianchi, M.S.A., Beçak, M.L. and Rabello, M.N. (1969) Chromosoma (Berl.) 26, 188-200.

Dutrillaux, B., Couturier, J., Mulleris, M., Lombard, M. and Chauvier, G. (1981) La Recherche 127, 1246-1255.

Hoge, A.R., Belluomini, H.E., Schreiber, G. and Penha, A.M. (1959) Mem. Inst. Butantan 29, 17-88.
Howell and Black (1980) Experientia 36, 1014-1015.
Jones, K.W. and Singh, L. (1985) TIG 1, 55-61.
Kiel-Metzger, K., Warren, G., Wilson, G.N. and Erickson, R.P. (1985) N. Engl. J. Med. 313, 242-245.
Kent, M.G., Elliston, K.O., Shroeder, W., Guise, W. and Wachtel, S.S. (1988) Cytogenet. Cell. Genet. 48, 99-102.
Kobel, H.R. (1962) Experientia 81, 173-174.
Lima-de-Faria, A. (1983) Molecular evolution and organization of the chromosome. Elsevier, Amsterdam, New York, Oxford.
McLaren, A., Simpson, E., Tomonari, P., Clander, H., Hogg (1984) Nature 312, 552-555.
Mengden, G.A. and Stock, A.D. (1980) Chromosoma (Berl.) 79, 53-64.
Mengden, G.A. (1981) Chromosoma (Berl.) 83, 275-287.
Moreno, R., Navarro, J., Iturra, P. and Veloso, A.M. (1987) Braz. J. Genetics 3, 497-506.
Ohno, S. (1967) Sex chromosomes and sex linked genes, Springer Berlin Heidelberg New York.
Nakamura, D., Wachtel, S.S., Lance, V. and Beçak, W. (1987) Proc. R. Soc. Lond. 232, 159-180.
Ray-Chaudhuri, S.P., Singh, L. and Sharma, T. (1970) Cytogenetics 9, 410-423.
Ray-Chaudhuri, S.P., Singh, L. and Sharma, T. (1971) Chromosoma (Berl.) 33, 239-251.
Ray-Chaudhuri,S.P. and Singh,L. (1972) The Nucleus 15, 200-210.
Singh, L. (1972b) Experientia (Basel) 28, 95-97.
Singh, L. (1974) Herpetologica 30(4), 419-429.
Singh, L., Purdom, I.F. and Jones, K.W. (1976) Chromosoma (Berl.) 59, 43-62.
Singh, L. and Jones, K.W. (1982) Cell 28, 205-216.
Singh, L., Phillips, A. and Jones, K.W. (1984) Cell 36, 111-120.
Singh, L., Purdom, I.F. and Jones, K.W. (1979a) Chromosoma (Berl.) 71, 167-181.
Singh, L., Purdom, I.F. and Jones, K.W. (1980) Chromosoma (Berl.) 79, 137-157.
Singh, L., Purdom, I.F. and Jones, K.W. (1981) Cold Spring Harb. Symp. Quant. Biol. 45, 805-813.
Singh, L., Ray-Chaudhuri, S.P., Majumdar, K., Purdom, I.F. and Jones, K.W. (1979b) Chromosoma (Berl.) 73, 93-108.
Singh, L., Sharma, T. and Ray-Chaudhuri, S.P. (1968a) Cytogenetics 7, 161-168.
Singh, L., Sharma, T. and Ray-Chaudhuri, S.P. (1968b) Experientia 24, 79-80.
Singh, L., Sharma, T. and Ray-Chaudhuri, S.P. (1970) Chromosoma (Berl.) 31, 386-391.
Sumner, A.T. (1972) Exp. Cell Res. 75, 304-306.
Sympson, E., Chandler, P., Goulmy, E., Disteche, C.M., Ferguson-Smith, M.A. and Page, D.C. (1987) Nature 326, 876-878.
Tartof, K.D. (1988) Genetics 120, 1-6.
Wachtel, S.S., Ohno, S., Koo, G.C. and Boyse, E.A. (1975) Nature 257, 235-236.

Cytogenetics of Amphibians and Reptiles
Advances in Life Sciences

CYTOGENETIC STUDIES IN LACERTID LIZARDS

L.Kupriyanova

Zoological Institute the USSR Academy of Sciences 199034 Leningrad

SUMMARY:Cytogenetic studies of the nucleolar organizer region and the sex determining system in 11 species from different genera of Lacertidae have revealed ancient origin of genera Takydromus and Eremias and heterogeneity of genera Lacerta,Archaeolacerta and Sg Zootoca. *Lacerta* (Sg Zootoca) *vivipara* Jaquin may be a composite group including several forms.

Cytogenetic studies of the karyologically uniform Lacertidae can contribute to understanding mechanism and trends of genome evolution, systematics and phylogeny.

Our data on the distribution of the nucleolar organizer region (NOR) in some species of Lacerta (Lacerta s.str.,Archaeo lacerta,Zootoca), Takydromus and Eremias argue that the character is useful in systematics and phylogeny. Localization of the NOR on one of the smallest pairs of chromosomes in *T.amurensis* confirms that the NOR is located on micro- or small chromosomes in ancient genera (Takydromus,Gallotia)(Olmo et.al.,1987). Localization of the NOR in two small chromosomes (approximately the 15th or 17th pair) in three species of Eremias (*E.arguta*, *E.velox*,*E.przewalskii*) suggests that this is also an ancient genus (Fig.1A). Our data on the position of the NOR in several species of the Lacerta (Lacerta s.str.,Archaeolacerta, Zootoca) have revealed heterogeneity,which accords with studies on other lacertid species (Olmo et.al.,1987). *Lacerta agilis* and *L.strigata* are included in the advanced Lacerta s.str.,the NOR bearing chromosomes being the 5th or 6th pair. *Lacerta saxicola*,*L.caucasica*,*L.dahli* and *L.mixta* are included in the Lacerta ,part11,Archaeolacerta (Arnold,1973), which itself seems to be heterogenous. Here the 4th or 5th pair are the NOR bearing chromosomes.

Lacerta (Zootoca) vivipara Jaquin appears to be unique for its sex chromosomes and NOR position the 11th or 12th pair are the NOR bearing chromosomes (Fig.1D-G).

Studies of NOR localization reveal different trends of chromosomal evolution in the Lacertidae because definite group of chromosomes are included in this process in the different genera or subgenera.

An other interesting point is the sex determining system. Many lacertid species from different genera possess two sex (WZ) chromosomes belong to 13th or 18 pairs. Only L.(Z)vivipara possesses multiple sex chromosome system with three sex chromosomes (W may belong to 4th or 6th pair, Z_1 to the 11 pair, Z_2 is the 15th pair, probably one of the Z chromosomes belongs to the 13 pair like this in other lacertids).

This $WZ_1Z_2/Z_1Z_1Z_2Z_2$ system in Lacerta (Sg Zootoca) vivipara may confirm its separation from the Lacerta and NOR localization may argue its similarity with Podarcis.

L.(Z)vivipara has various sex chromosome configurations within its range. K.Oguma (1934) first described 35 acrocentric (A) chromosomes in females and 36 (A) chromosomes in males of L.vivipara from Sakhalin,ZO/ZZ system.A homomorphic (2n=36 A) karyotype was found in both males and females from several populations in Europe of this species (Margot,1946;Brink,1959;Orlova and Orlov,1969). An other heteromorphic karyotype with 2n =35 (1SV+34A) in female and 36A in male,NF=36 was found.Their interpretation was that the W submetacentric (SV)chromosome (4th or 6th pair) had derived via centric fusion of Z_1 and Z_2 giving a $WZ_1Z_2/Z_1Z_1Z_2Z_2$ system (Chevalier,1969;Chevalier et.al., 1979;Belcheva et.al.,1986).

Therefore this species from different localities throught the USSR has been investigated.Karyological investigations of L.vivipara from Sakhalin Isl. supported the diploid numbers described by K.Oguma (2n=35A♀/36A♂).However unpaired acrocentric W chromosome (4th or 6th pair) and two intermediate small unpaired acrocentric Z_1 of pair 11 and Z_2 of pair 15 were revealed.

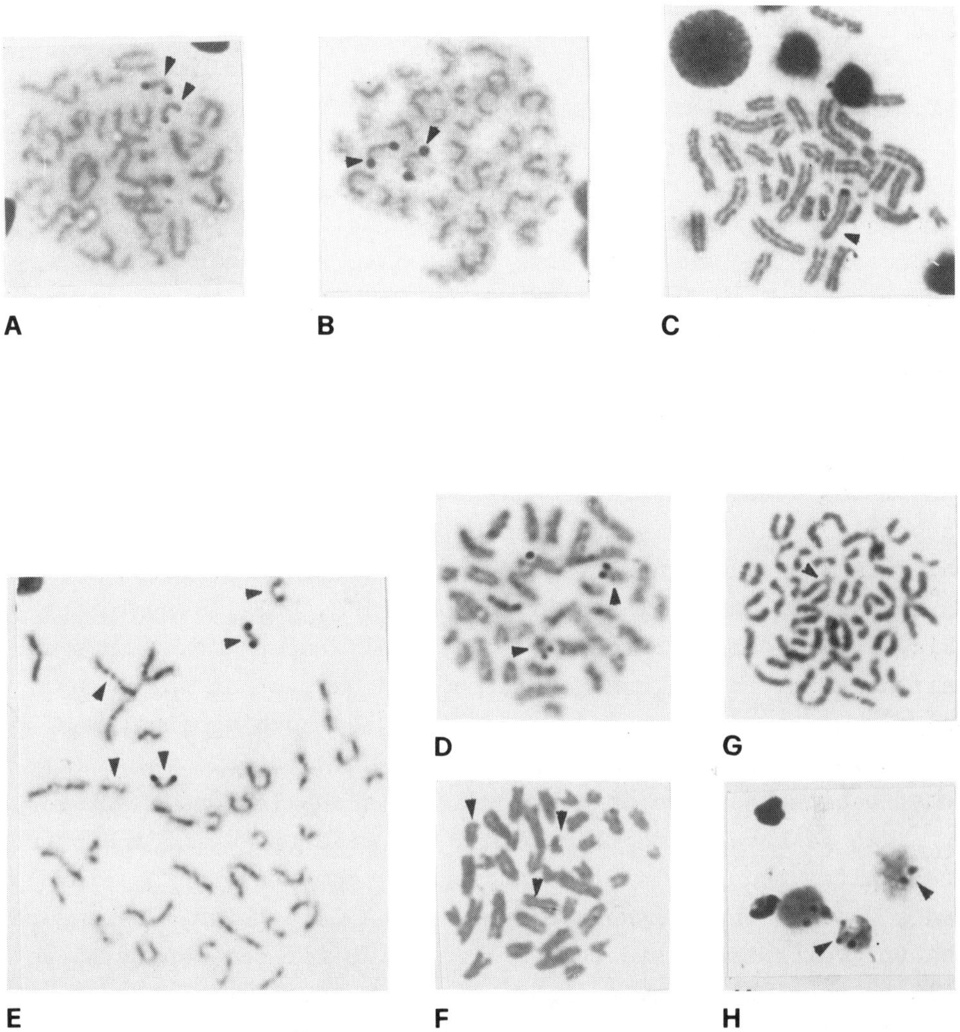

<u>Figure 1</u>: Metaphases of somatic cells (A) <u>Eremias</u> <u>arguta</u>, (B) <u>Lacerta</u> <u>agilis</u>, (C) <u>L.dahli</u>, (D) <u>L.(Zootoca)</u> <u>vivipara</u> (♂ 2n=36), (F) <u>L</u>.(Z).<u>vivipara</u> (♀ 2n=34A+1SV),form B, (E,G) <u>L</u>.(<u>Z</u>) <u>vivipara</u> (♀ 2n=35A), formA,(H) <u>L</u>.(<u>Z</u>) <u>vivipara</u> two interphase nucleoli.Arrows point to NOR bearing chromosome and sex chromosomes.

This W chromosome might have arisen via a tandem fusion of Z_1 and Z_2 or via a pericentric inversion of submetacentric W,WZ_1Z_2/ $Z_1Z_1Z_2Z_2$ system (Kupriyanova,1986;Kupriyanova and Rudi in press)

Some inter- and intrapopulational variations in sex chromosome morphology are known in different species.However we have found no variations or polymorphism in the morphology of W chromosome although short arms on it have sometimes been observed (Fig.1G).We named L.(Z) vivipara with acrocentric W chromosome form A and with submetacentric W chromosome form B (Kupriyanova and Rudi, in press).Form A with acrocentric W chromosome has been discovered in 9 populations throughout the USSR from one locality in the southern slopes, then from the northern slopes of the Eastern Carpathian up to Sakhalin.Form B with submetacentric W chromosome has been observed by us in two populations from the southern slopes of the Western, near Zakopane,Poland, and of Eastern, near v.Shirokii lug, USSR, Carpathians (Fig.1E,F).The form B inhabite the main European mountain system:the Alps, the Balkan Mountains, the Carpathian which may suggest a unity of these faunes.We can say nothing about the distribution range of the form with the homomorphic sex chromosomes because we have not observed it in populations examined.

As we have found that form A has wide range and inhabits Urals/Altai mountain system (ancient as compared to the Alps) and a zone of the southern slopes of the Carpathian, Chernogora region (which is considered as a refugium for some animals) we believe that this form A might have been the initial one (Kupriyanova and Rudi, in press). This interpretation means that the first step in the course of evolution of L.(Z) vivipara (2n =36A) may be a tandem fusion of Z_1 and Z_2 chromosomes resulting in an appearance of acrocentric W chromosome (form A),followed by a pericentric inversion in acrocentric W chromosome (form B).

One should not be rule out a possibility that the rearangements may have occured during a formation of the Alps system accompanied by seismic and some other factors.Various chromosomal genome mutations (autopolyploidy, inversions, B chromosomes) which have been observed in different mammalian and lizard species from the Balcan Mountains, the Caucasus, the Tien Shan (Vorontsov and Lyapounova ,1984;Kupriyanova and Rudi in press) support an association between geological activity and chromosomal change.

Some chromosomal rearangements are known to be associated with a speciation. Translocations of unique sex chromosomes in L.(Z) vivipara may have produced (1) an unique gene combination, which may in turn be related to its ovoviviparity (in both forms by our observations) and a reinforced control for sex determination and an isolation. These forms with different karyotype structure and with separate distribution ranges suggest that L.(Z) vivipara may be a composite group and the taxonomic status of the forms should be checked.Mating between them was not successful during one season, but the experiment should be repeated.Also there should be additional cytogenetical investigations of L.(Z) vivipara from Western and Northern part of Europe,especially from the Pyreneen,the Alps.Special emphasis should be placed on revealing of these forms there and of their para- or sympatric areas.

REFERENCES

Arnold,E.N. (1973) Bull.Brit.Mus.Nat.Hist.Zool. 8,291-366.

Belcheva,R.,Biserkov,V.,Ivlieva,H.,Beschkov,V.,Petkov,P. (1986) Cytologia 3, 567-570.

Brink,G.M. van (1959) Chromosoma 1, 1-72.

Chevalier, M. (1969) C.R.Acad.Sci. 16, 2098-2100.

Chevalier,M.,Dufaure,J.P.,Lecher,P. (1979) Genetica 1,11-18.

Kupriyanova,L.A. (1986) In:Systematic and Ecology of Amphibians a nd Reptiles,vol.157 (N.Ananieva and L.Borkin,Eds),Naika,Leningrad,86-100.

Kupriyanova ,L.A. and Rudi,E.R.Zool.J. (in press).

Margot, A (1946) Rev.suisse Zool.28,556-596.

Oguma,K. (1934) Arch.Biol.vol.45,27-46.

Olmo,E.,Odierna,G.,Capriglione,T. (1987) Proc.4th ordin. gen. Meet.Soc.Herp.Europ.Nijmegen, 303-306.

Orlova,V. and Orlov,V.(1969) Zool.J.,7,1056-1060.

Vorontsov,N. and Lyapunova,E. (1984) In:Chromosomes today, vol. 8 (G.Allen & Unwin,Eds),pp.179-294.

Cytogenetics of Amphibians and Reptiles
Advances in Life Sciences

GENOMIC INTRASPECIFIC VARIATIONS IN SOME POPULATIONS OF *Podarcis s. sicula* Raf.

A. Cardone, T. Capriglione, G. Odierna, C.A. Redi*, S. Garagna*

Dipartimento di Biologia Evolutiva e Comparata, Universita´ di Napoli, Napoli (Italy); * Dipartimento di Biologia Animale, Universita´ di Pavia, Pavia (Italy).

SUMMARY: Intraspecific variations in the chromosomal C-banding pattern of different *Podarcis s. sicula* populations captured in different areas of Campania, were observed. The C-banding technique support the presence of two different band patterns: the first, characterized by centromeric bands, is more wide-spride in the Campania region; the second, characterized by paracentromeric bands, is confined only to some areas of Cilento. Quantitative heterochromatin by flow cytometer do not show statistically significant variation in DNA-content or in particular DNA fractions between two cytotypes. In addition, two fractions of highly repetitive DNA were isolated from the genome of the lizard *P. s. sicula*. One fraction, that appears after Taq I enzyme digestion, is present in all the populations studied and is mainly localized on the centromere regions. The other fraction, that were visualized by Hind III digestion, seems to be interspersed along the chromosomes.

INTRODUCTION

Some organisms, mainly mammals, show a high intraspecific variability in the content of constitutive heterochromatin and of highly repeated DNA associated with it. It has been suggested that this variability is involved in the process of diversification among populations, which is the basis of speciation phenomena (Deaven et al.,1977; Patton and Sherwood,1982).

In the lizard *Podarcis s. sicula*, intraspecific chromosome variation has been observed in C-banding. Olmo et al. (1986) have evidenced two different cytotypes which seem to be found in different geographical areas: a cytotype with centromeric C-bands and quite a large amount of heterochromatin (25%), and another showing paracentromeric C-bands and a much lower amount of heterochromatin (12%) (Olmo et al.,1988).

Histochemical and biomolecular data indicate the presence of particular DNA fractions in *P. s. sicula* constitutive heterochromatin. In fact, in both cytotypes of this species, the C-bands intensely stain with DAPI, a fluorochrome specific for DNA rich in adenine-thymine (Olmo et al.,1988).

In addition, from the lizard population with centromeric C-banding, Capriglione et al. (1989) have isolated and identified a satellite DNA which is localized at the centromeric C-bands level.

In the present paper, the study of variations in the C-banding pattern was extended to various *P. s. sicula* populations coming from different areas of Campania. Moreover, observations were conducted to learn whether variation in the amount of constitutive heterochromatin occurs concomitantly with variation in DNA-content or in particular fractions of it. Afterwards, the content of the DNA and of the DNA staining with DAPI was evaluated in specimens with either centromeric or paracentromeric C-banding. The genome of specimens with both centromeric and paracentromeric banding was also studied using restriction enzymes.

MATERIALS AND METHODS

Animals. The specimens studied belonged to populations of Podarcis s. sicula captured in different areas of Campania, as is show in Fig. 1.

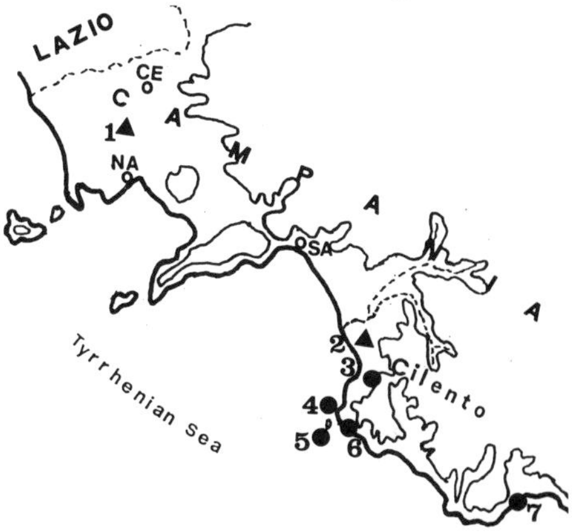

Fig.1- Map of geographic distribution of different P. s. sicula cytotypes characterized by different localizations of C-bands. ▲cytotype with centromeric bands; ● cytotype with paracentromeric bands. 1 Regi Lagni; 2 Paestum; 3 Prignano Cilento; 4 Punta Licosa; 5 Isola di Licosa; 6 Ogliastro Marina; 7 Scario.

Cytological preparation. Mitotic slides were prepared according to Olmo et al. (1986). C-banding was performed with Sumner´s $Ba(OH)_2$ method (1972) modified by Odierna et al. (1985).

Quantitative measurements. The percentage of heterochromatin was obtained by measuring the ratio of the C-bands area to the overall chromosome area by Zeiss-Kontron IBAS 1 image analyser. The DNA-content was evaluated on blood and liver preparations stained with propidium iodide (50 μg/ml), an intercalating stechiometric aspecific dye into the DNA double strand. Variation in the content of AT bases was evaluated by staining the same samples with DAPI (5 μg/ml), a fluorocrome specific for AT-rich DNA. Measurements were performed using two flow cytometers, PAS 2 (Partec) and FACStar (Becton Dickinson).

Restriction enzyme analysis, Southern DNA blot hybridization experiments. DNA extracted from liver and testes, according to Jeffrey and Flavell (1977), was analyzed by the restriction enzyme digestion method. All restriction enzymes were purchased from Bethesda Research Laboratories and used according to the manufacturer´s reccomendation. Digested DNAs were electrophoresed under costant voltage (60-80 V) on horizontal agarose gels (1.2% w/v) in TBE (0.089M Tris-borate, 0.089M Boric acid,0.002M EDTA) running buffer and visualized on a UV transilluminator after staining with ethidium bromide (1 μg/ml in TBE). After electrophoresis, fractionated DNAs were transferred onto nitrocellulose filters according to Southern (1975). For hybridization, DNA on filters was annealed to double-stranded probes labelled with ^{32}P by nick translation according to the procedure described by Capriglione et al.(1989). The restriction enzyme analysis was conducted also in metaphase chromosomes preparated with the above mentioned technique. Chromosomes were digested for 12 hrs with 100 Units of the enzyme as described by Capriglione et al. (1989).

RESULTS AND DISCUSSION

C-banding results on specimens from different populations support the presence of two band patterns: a pattern with centromeric bands, which is observed essentially in lizards wide-spread in Campania; and another one showing paracentromeric bands, which would be found only in some areas of Cilento. Quantitative evaluation of constitutive heterochromatin, moreover, confirms that is more abundant in specimens from populations with centromeric bands than in those from populations with paracentromeric bands (Tab.I). However, there are only slight, statistically insignificant differences in DNA-content and DAPI stainable DNA among specimens from different populations (Tab.II).

Genome analysis by restriction enzymes showed same homogeneity among the various Podarcis s. sicula populations.

As already mentioned, from genome of P. s. sicula populations with centromeric band patterns, Capriglione et al. (1989) have isolated a satellite DNA by treatment with the Taq I enzyme. The smaller repeated unit of this satellite is about 200 bp, and in situ hybridization has shown that it is localized at the centromeric level.

Treatment of the genomic DNA with Taq I evidenced the same band pattern in all of the specimens studied by us (Fig.2). This is indicative of the presence of satellite DNA resembling the one isolated by Capriglione et al. (1989). This result was confirmed by hybridizing the genomic DNA of two P. s. sicula populations with different band patterns to the Taq I satellite DNA (pLCS1) cloned by Capriglione et al. (1989). In fact, the autoradiography shows the same hybridization pattern in both populations (Fig.3).

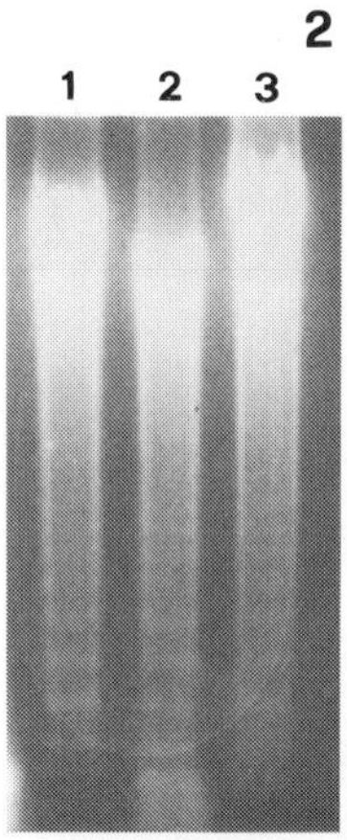

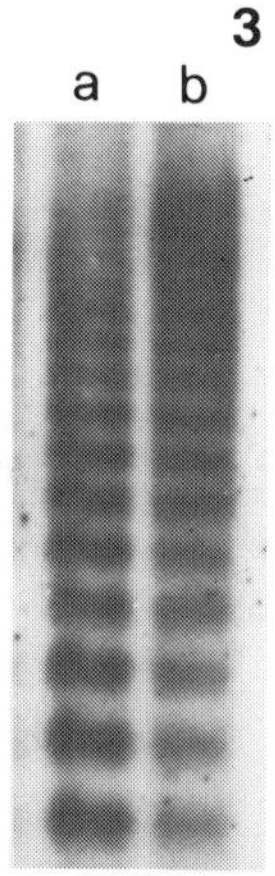

Fig.2- Ethidium bromide stained agarose gel of Taq I-digested genomic DNA from three different P. s. sicula populations. Lane 1:Regi Lagni; lane 2:Punta Licosa; lane 3:Prignano Cilento.

Fig.3- Hybridization of pLCS1 to Taq I digested genomic DNAs from two P. s. sicula populations:a) Punta Licosa,b) Regi Lagni.

P. s. sicula populations	C-banded chromosome number	Heterochromatin (%)	HR-DNA (%)
Scario	26	15.4	
Isola di Licosa	24	12.4	22
Punta Licosa	26	18.4	
Paestum	26	15.1	
Regi Lagni	28	24.5	22

Table I. Variations of heterochromatin amounts in different *P. s. sicula* populations measured by image analyzer. Populations from Cilento show a lower percentage of heterochromatin and fewer banded chromosomes.

P. s. sicula populations	PROPIDIUM thym/blood		DNA-Content (pg)		DAPI thym/blood
	PARTEC	FACSTAR	PARTEC	FACSTAR	PARTEC
Punta Licosa	2.17	1.85	3.13	3.67	2.21
Ogliastro Marina	2.09	1.83	3.25	3.71	2.10
Regi Lagni	2.09	1.80	3.25	3.77	2.29

Table II. Data on the DNA content and the DAPI positive DNA amount in three populations of *P. s. sicula*. Variations are not statistically significant. Thym/blood= ratio between lizard blood cells value and mouse thymocyte value as a standard.

The genome of P. s. sicula shows another highly repeated sequence, which can be isolated by digestion with Hind III. Like the Taq I satellite, it displays the same band pattern in all the specimens from the different populations (Fig.4). Howewer, this sequence does not seem to bind C-banded heterochromatin, but might be uniformly distibuted along the chromosomes. In fact, though *in situ* hybridization tests were not made, treatment of metaphase chromosomes with Hind III evidenced several sites responsive to this enzyme along the chromosome arms (Fig.5).

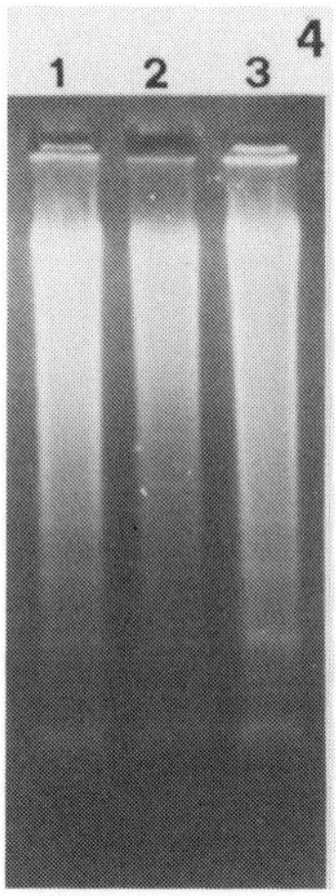

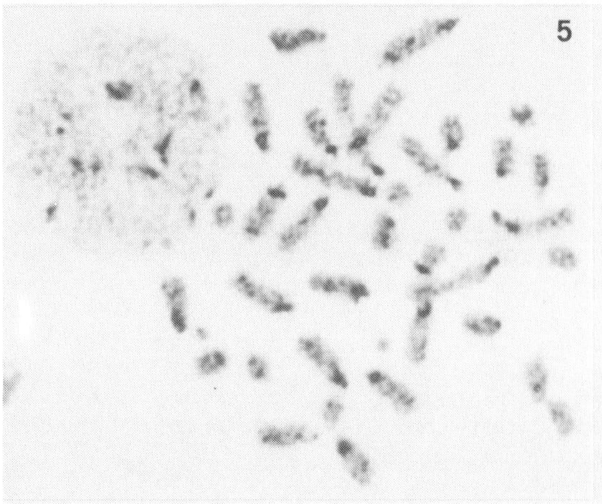

Fig.4- Ethidium bromide stained agarose gel of Hind III-digested genomic DNA from three different P. s. sicula populations. Lane 1:Regi Lagni; lane 2:Punta Licosa; lane 3:Prignano Cilento.
Fig.5- Metaphasic chromosomes of P. s. sicula treated with Hind III.

CONCLUSIONS

Our results, though being preliminary, seem to suggest that the different C-band patterns observed in the various *P. s. sicula* populations do not depend, or depend only in a small part, on differences in DNA. In fact, no significant differences were observed between the various populations either in DNA-content, or in the amount of DAPI-stainable DNA; in addition, a marked homogeneity was detected by means of restriction enzymes.

Therefore, the difference in C-banding and heterochromatin amount are more likely due either to differences in protein content or in chromatin structure at the level of the C-bands.

REFERENCES

Capriglione T.,Olmo E.,Odierna G.,Smith D.I.,Miller O.J. (1989). Genetica (in press).

Deaven L.L., Vidal-Rioja J., Jett J.H., Hsu T.C. (1977). Cytogenet. Cell Genet. 19, 241-249.

Jeffreys A.J., Flavell R.A. (1977). Cell 12, 429-439.

Odierna G., Olmo E., Cobror O. (1985). Experientia 41, 944-946.

Olmo E., Odierna G., Cobror O., 1986). Genetica 71, 63-74.

Olmo E., Odierna G., Capriglione T., Merciai B.M. (1988). Comp. Biochem. Physiol. 89B, 1-4.

Patton J.L., Sherwood S.W. (1982). Chromosoma 85, 149-162.

Southern E.M. (1975). J. Mol. Biol. 98, 503-517.

Sumner A.T. (1972). Exp. Cell. 75, 304-306.

Cytogenetics of Amphibians and Reptiles
Advances in Life Sciences

KARYOTYPES OF SOME SPECIES OF JAPANESE COLUBRID SNAKES

Michihisa Toriba

Japan Snake Institute, Yabuzuka-honmachi, Nitta-gun, Gunma, 379-23 Japan

SUMMARY: The karyotypes of five species of Japanese colubrid snakes are described. Those of *Amphiesma vibakari* and *A. pryeri* are similar each other, and different from that reported previously. The karyotype of *E. conspicillata* is similar to those of other species of *Elaphe*. Two species of *Dinodon* showed large intrageneric variation.

Recently, cytogenetic study on the snakes is rather poor in Japan, in spite of pioneer work by Nakamura (1935). In the course of study on the chromosomes of snakes, several interesting results were obtained on five species of colubrid snakes, including some subspecies.

MATERIALS AND METHODS

The following specimens were used, which includes some from outside Japan for comparison. *Elaphe conspicillata* (two females from Gunma and Miyagi Prefectures); *Amphiesma vibakari vibakari* (two females from Gunma); *A. v. ruthveni*(a female from USSR); *A. pryeri pryeri* (a female from Okinawa Is.); *Dinodon rufozonatus*

rufozonatus (a female from Taiwan); D. r. walli (a female from Ishigaki Is.); D. semicarinatus (one pair from Amami-Oshima Is.).

Chromosomes were observed from bone marrow of ribs. Colchicin was injected to the snakes 7-8 hours before sacrifice. The details of the preparation was given by Yosida and Toriba (1986).

RESULTS AND DISCUSSION

Elaphe conspicillata: The diploid number was 36 consisting of 16 macro and 20 microchromosomes (Fig. 1). The number and arrangement of chromosomes are similar to those of two other congeners, E. climacophora and E. quadrivirgata, in Japan.

Amphiesma vibakari: All the specimens showed the same karyotypes consisting of 14 macro and 22 microchromosomes, resulting in the diploid number of 36 (Fig. 2). The third pair has secondary constriction on the tip of longer arm, although it is not clear in Fig. 2. The karyotype observed here is entirely different from that reported by Itoh et al. (1970). They have examined a female Natrix (=Amphiesma) vibakari whose origin was not stated. Because the figure presented by them was identical with that of genus Elaphe, I have considered that they have examined E. conspicillata under the misidentified name, N. vibakari. In eastern area of Honshu Is., Japan, these two species have the same vernacular name "Jimuguri" and are usually confused by the people. Such a situation might have led them to the mistake.

Amphiesma pryeri: The karyotype is similar to that of A. vibakari (Fig. 3). In this figure, secondary constriction is clearly visible. This evidence supports the view of Malnate (1962) who suggested their close relationship. On the other hand, the karyotypes of these two species are very different from that of A. stolata reported by Singh (1972). To demonstrate the difference, their idiograms of macrochromosomes are shown in Fig. 4. At present it is difficult to explain the

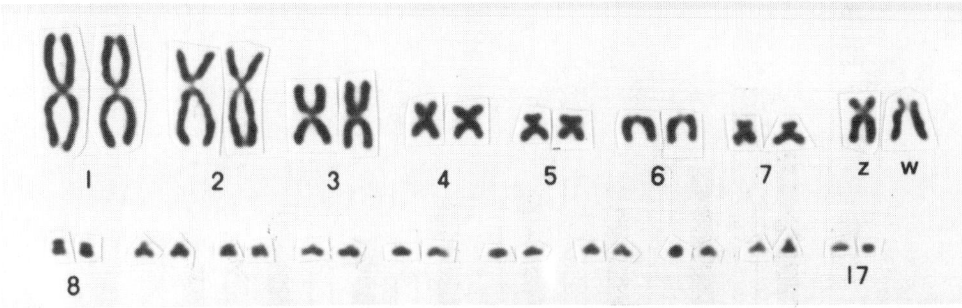

Figure 1: Karyotype of a female Elaphe conspicillata from Nitta, Gunma Prefecture, Japan.

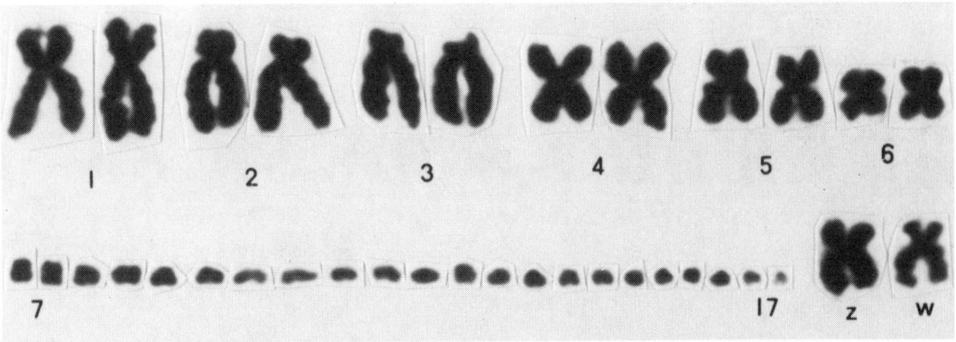

Figure 2: Karyotype of a female Amphiesma v. vibakari from Kiryu, Gunma.

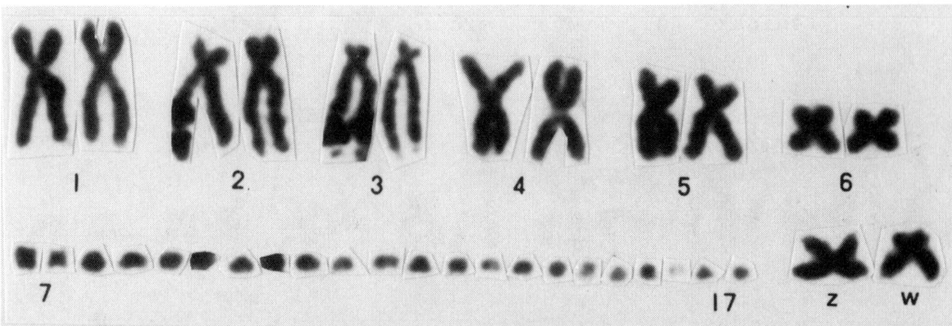

Figure 3: Karyotype of a female Amphiesma p. pryeri from Okinawa Is., Japan.

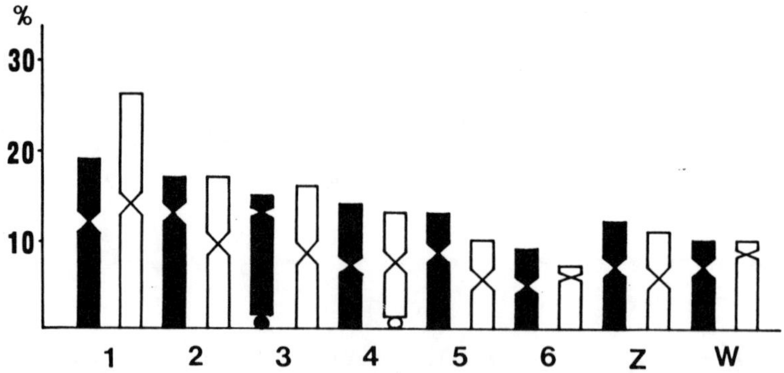

Figure 4: Idiograms of macrochromosomes of <u>Amphiesma vibakari</u> (solid) and <u>A</u>. <u>stolata</u> (hollow).

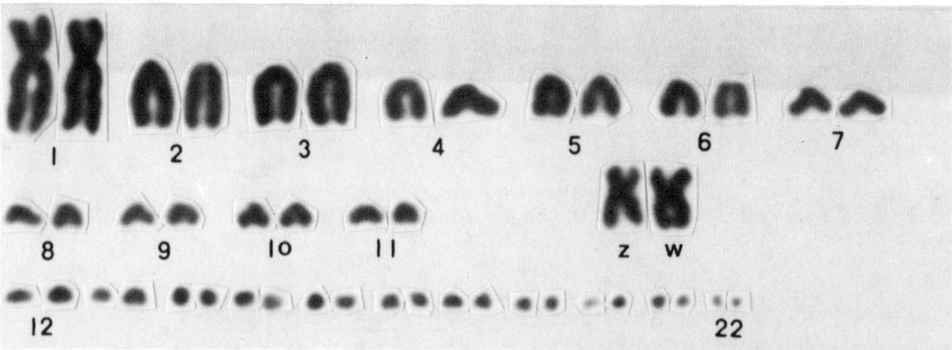

Figure 5: Karyotype of a female <u>Dinodon</u> <u>r</u>. <u>walli</u> from Ishigaki Is., Yaeyama group, Ryukyu Arch.

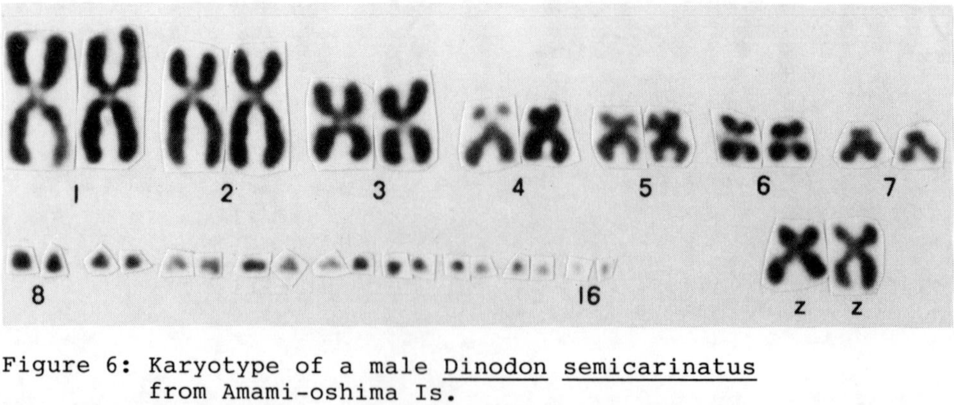

Figure 6: Karyotype of a male <u>Dinodon</u> <u>semicarinatus</u> from Amami-oshima Is.

difference. Future study on other members of the genus will clarify it.

Dinodon rufozonatus: The two subspecies have similar karyotypes and the diploid number was 46 consisting of 24 macro and 22 microchromosomes (Fig. 5). However, the separation between macro and micro is not clear. Therefore, Nakamura (1935) considered the macro as only 8 pairs. Only largest pair and sex chromosomes are biarmed and the rest is uniarmed. In this respect, Nakamura was incorrect, because he recognized only one pair of V-shaped chromosomes.

On some slides of D. r. walli, C-banding stain was also made. Mainly centromeric areas and tip of arms were well stained in the autosomes. While in sex chromosomes, W was well stained. Thus determined W has slightly higher arm ratio than Z.

Dinodon semicarinatus: The diploid number was 34 consisting of 16 macro and 18 microchromosomes (Fig. 6). Most macrochromosomes are biarmed. And a secondary constriction was observed in the shorter arm of 4th pair. W is metacentric and has slightly lower arm ratio than Z. Although the difference in the karyotypes between the two species of the genus Dinodon examined here is striking, number of arms is identical and the occurrence of Robertsonian change was suggested.

Acknowledgements: I am grateful to Vladimir Spitsin and Sergei Kudrjavtsev of Moscow Zoo, and Yasuji Hokama and Akira Nakachi of Japan Snake Institute for collecting snakes.

REFERENCES

Itoh, M., Sasaki, M. and Makino, S. (1970) Jpn. J. Genet. 45, 121-128.
Malnate, E. V. (1962) Proc. Acad. Nat. Sci. Philadelphia 114, 251-299.
Nakamura, K. (1935) Mem. Coll. Sci. Kyoto Imp. Univ. B10, 361-402.
Singh, L. (1972) Chromosoma 38, 185-236.
Yosida, T.H. and Toriba, M. (1986) Proc. Jpn. Acad. 62B, 13-16.

Cytogenetics of Amphibians and Reptiles
Advances in Life Sciences

KARYOTYPES OF GULF OF CALIFORNIA INSULAR RATTLESNAKES (VIPERIDAE: *CROTALUS*) COMPARED TO THOSE OF PENINSULAR SISTER TAXA

S. Stewart, D. Morafka; and A. Stock

Department of Biology, California State University, Dominguez Hills, Carson, California 90747 U.S.A.; Genetics Network, University of Nevada School of Medicine, 888 Willow St., Reno, Nevada 89502 U.S.A.

SUMMARY: Three rattlesnake (*Crotalus*) species of Baja California were compared to three possible relatives from islands of the Gulf of California, by means of G and C Banding, to determine whether karyotyping might provide data regarding evolutionary relationships. Banding revealed conservatism except in the W of the ZW chromosomes of the females.

Since at least 1973, it has been known that crotaline snakes of North America exhibit conservatism in the diploid number of their chromosomes (Zimmerman and Kilpatrick, 1973). Studies of diverse *Crotalus* (rattlesnake) species revealed no interspecific variation in the autosomes, although some differences were found to exist in the ZZ/ZW sex chromosomes.
The present investigation was initiated with the expectation that perhaps G and C Banding would reveal more about the relationship between *Crotalus* species than could be detected through simple karyotyping.

The Baja California region was selected because the distribution of Peninsular and Gulf of California island taxa provides an unusual opportunity to test the stability of the rattlesnake karyotype, both spatially and temporally. Tectonic events occurred approximately 13 MYBP (million years before the present), when the Cape of the Baja Peninsula broke away from the mainland because of the northwestern movement of the Pacific

Plate and the southeastern movement of the North American Plate. With the formation of a Proto-Gulf from these plate movements, populations of the resulting Peninsula, Cape, and some islands became isolated from those of the mainland.

Crotalus species account for four peninsular species, two named island species, and several additional subspecies (not including putative island species currently unnamed). Three additional forms occur on the opposite Sonoran Desert mainland. Three island species and three Peninsular and mainland putative sister taxa were selected for this study; (Table I).

Table I: Species selected for study

Island/Population	Possible Ancestor	Origin
Isla Santa Cruz		
Crotalus ruber	C. ruber	Peninsula
Isla Santa Catalina		
C. catalinensis	C. ruber, C. enyo	Peninsula
	C. scutulatus	Mainland Sonoran
Isla Tortuga		
C. tortugensis	C. ruber	Peninsula

METHOD

Live Crotalus specimens were collected by Morafka and Murphy in Baja California and California, in May 1988. (Table II).

Table II: Species, sex, and locality

Species	Sex	Locality	Country	Long./Lat.
C. tortugensis	♂	Isla Tortuga	Mex.	111°52'W/27°28'N
C. catalinensis	♀	Isla Catalina	Mex.	110°47'W/26°39'N
C. ruber	♂	Isla Santa Cruz	Mex.	111°43'W/25°17'N
C. ruber	♂	Riverside County	U.S.A.	117°21'W/33°59'N
C. enyo	♀ ♂	Bahia de los Angeles	Mex.	112°33'W/28°53'N
C. scutulatus	♂	Needles, California	U.S.A.	114°39'W/34°50'N

Protocol for tissue removal and culturing was according to the procedure of Kurth, et al. (manuscript). Under sterile conditions, small samples approximately (.5cc), of kidney, gonad, lung, heart, and dermal tissue were excised from each specimen and placed in culture. After good fibroblast growth was obtained, harvesting of cells and G Banding techniques were carried out for each species collected, using techniques of Stock (1984), and Kurth, et al. (manuscript). C Banding techniques were according to a procedure of Sumner (1972), modified by Bull (1978), and by Stock.

RESULTS

All six species that were G Banded revealed a diploid chromosome number of 36 with 16 macrochromosomes and 20 microchromosomes, and a NF of 52 (chromosomal arms). The morphology of the macrochromosomes, based on centromere position, included two pairs of metacentrics, four pairs of submetacentrics, and two pairs of telocentrics. The second largest pair of chromosomes displayed a large, lightly stained area around the centromere (G-C rich heterochromatin). The telomeric ends of all the macrochromosomes also contained light regions which also correspond to regions of heterochromatin (Fig. 1).

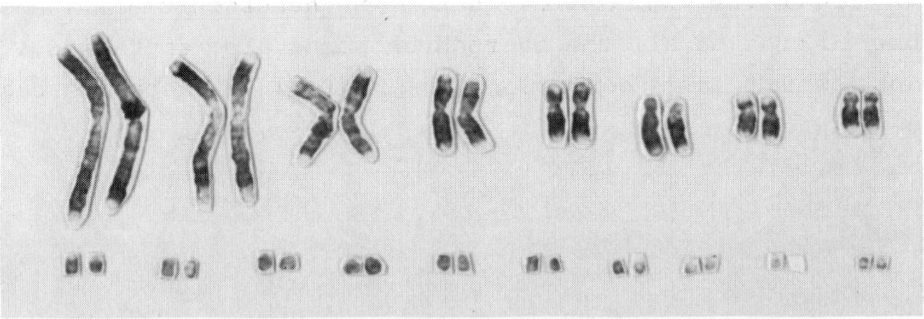

Fig. 1 G Banded chromosomes of C. tortugensis (male).

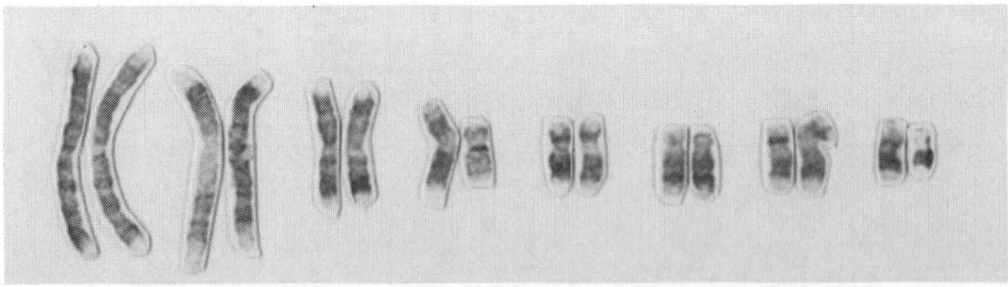

Fig. 2 G Banded macrochromosomes of C. catalinensis (female). Note the ZW sex chromosomes (fourth pair).

The ZZ/ZW sex chromosomes are of the most interest. The female exhibits the ZW chromosome in which differences exist between some species in the W chromosome. This was true in our study as well as that done by Kurth, et al. (manuscript). The G Band pattern of the W chromosome of C. catalinensis consisted almost entirely of lightly stained G-C rich chromatin, with only one dark band at the centromere (Fig. 2).

By contrast, C Banding produced dark bands particularly at the telomeric ends of the macrochromosomes and at the position of the centromere.

CONCLUSION

Despite their diversity, the Crotalus species of Baja California, chosen for this study, demonstrated autosomal stability in the diploid chromosome number of 36 and in the NF number of 52. This agrees with previous studies (Monroe, 1962; Baker, et al., 1972; Zimmerman and Kilpatrick, 1973; Kurth, et al., manuscript). The autosomal G and C Banding was consistent among all of the six species studied. These allopathic island populations, C. tortugensis, C. catalinensis, and C. ruber, exhibit no autosomal autapomorphies, nor does the trio share any unique synapomorphies in common. Despite their considerable spatial and temporal isolation, if insular species were actually sequestered by the original tectonic events which created the islands 3-6 MYBP, according to Murphy, the temporal factor is indeed considerable.

Relative to the autosomal chromosomes and the ZZ male chromosomes, the variability evident in the W chromosome of the ZW pair of sex chromosomes of the two females studied, C. catalinensis and C. enyo, will need further investigation to document the degree of W chromosome change and the evolutionary significance, if any.

Rattlesnakes are good candidates for founder effects. They are known to be able and enduring swimmers (Klauber, 1972). A bottleneck may have occurred in Isla Catalina and Isla Santa

Cruz with the rising of the Gulf waters which inundated the area up to the San Gorgonio Barrier and formed the Bouse Embayment that reached the present Parker Dam (Murphy, 1983; Lamb, 1989). The smaller number of members of the gene pool might make a chromosome difference more likely to persist and multiply in populations.

Continued research is necessary to establish a robust genealogy for all the rattlesnakes. It is most important to obtain females of each species in order to evaluate the ZW chromosomes, since these appear to be the most informative by virtue of their polymorphisms.

REFERENCES

Bull, J.J. 1978. Sex chromosome differentiation: an intermediate stage in a lizard. Can. J. Genet. Cytol. 20:205-209.

Baker, R.J., G.A. Mendgen, J.J. Bull. 1972. Karyotypic studies of thirty eight species of North American snakes. Copeia 2:257-265.

Hausback, B.P., V.A. Frizzell, Jr. Ed. 1984. Cenozoic volcanic and tectonic evolution of Baja California Sur, Mexico. Geol. of the Baja California Peninsula. Pacific Section Soc. of Econ. Paleon. and Mineral. Pub. Los Angeles, CA. 219-236.

Klauber, L.M. 1972. Rattlesnakes. Vol. 1. Univ. of Calif. Press. Berkeley.

Kurth, M.J., D.J. Morafka, A.D. Stock. (manuscript). Banded karyotypes from snakes: a technique adapted for both field and laboratory biopsy procedures.

Lamb, T., J.C. Avise, J.W. Gibbons. 1989. Phylogeographic patterns in mitochrondrial DNA of the desert tortoise (Xerobatess agassizi) and evolutionary relationships among the North American gopher tortoises. Evol. 43:76-87.

Monroe, J.E. 1962. Chromosomes of rattlesnakes. Herpe.17:217-220.

Murphy, R.W., C.B. Crabtree. 1985. Acta Zool. Mex. 9: Genetic relationships of the Santa Catalina island rattleless rattlesnake, Crotalus catalinensis (serpentes: viperidae).

Stock, A.D. 1984. The occurence of G-Bands in the mitotic chromosomes of the amphibian Xenopus meulleri. Genetica 64:225-228.

Sumner, A.T. 1972. A simple technique for demonstrating centromeric heterochromatin. Expl. Cell Res. 75:304-306.

Zimmerman, E.G., C.W. Kilpatrick. 1973. Karyology of North American crotaline snakes (family viperidae of the genera Agkistrtodon, Sisturus and Crotalus. Can. J. Genet. Cytol.: 389-395.

SUBJECT AND SPECIES INDEX